THE INTERNATIONAL COMMUNICATION OF TECHNOLOGY

SERIES ON INTERNATIONAL BUSINESS AND TRADE

SERIES EDITOR-IN-CHIEF
Khosrow Fatemi

THE INTERNATIONAL COMMUNICATION OF TECHNOLOGY
A Book of Readings

Edited by
Richard D. Robinson
University of Puget Sound

Taylor & Francis
New York • Philadelphia • Washington D.C. • London

T
10.5
.I53
1991

USA	Publishing Office:	Taylor & Francis New York Inc. 79 Madison Ave., New York, NY 10016-7892
	Sales Office:	Taylor & Francis Inc. 1900 Frost Road, Bristol, PA 19007-1598
UK		Taylor & Francis Ltd. 4 John St., London WC1N 2ET

THE INTERNATIONAL COMMUNICATION OF TECHNOLOGY: A Book of Readings

1 2 3 4 5 6 7 8 9 0   B R B R   9 8 7 6 5 4 3 2 1

A CIP catalog record for this book is available from the British Library.

Library of Congress Cataloging in Publication Data

The International communication of technology : a book of readings /
  Richard D. Robinson (editor).
      p.   cm.
  Includes bibliographical references.
  1. Communication of technical information. I. Robinson, Richard
D., 1921–
  T10.5.I53   1990
  601.4—dc20   90-11088
  ISBN 0-8448-1655-8   CIP
  ISSN 1052-9160

Contents

Foreword

This is the first volume of the *Series on International Business and Trade,* to be published by Taylor & Francis. We are very pleased that Professor Richard Robinson, a true pioneer in the study of international business, agreed to edit this initial publication, which will, indeed, set a fine example for others to follow.

The *Series on International Business and Trade* is a manifestation of the increasing significance of international business and global trade in the contemporary world. It is a reflection of the "new international economic order," one that will dominate the economic events of the 1990s and beyond. Unlike the short-lived attempts of the 1970s to create artificially a new international economic order, mainly in response to a series of increases in oil prices, the current metamorphosis is autonomous and is aimed at accommodating market-dictated changes. Probably the most striking characteristic of this new international economic order is the greater interdependency that it has brought about among different countries of the world. And nowhere is this evolutionary transformation better manifested than in international trade. For example, during the 1980s, the volume of global trade increased by 140 percent—twice the rate of growth in global production.

National policymakers and corporate executives, not to mention the academic community, may have been slow in realizing and reacting to this global economic restructuring, but finally a definite trend—acknowledging this new reality—has been established. It is my expectation that this trend will continue, indeed accelerate, and the study of international business and trade issues will continue to dominate the economic and political scene. Instead of the bipolarity issues of the recent past, the dominant themes of the future will include such international economic concerns as Europe–1992, the U.S.–Canada Free Trade Agreement, and the economic expansion of the Far East. Furthermore, it is my hope that the *Series on International Business and Trade* will play a major role in bringing about a greater awareness of the significance of international issues in today's business, and for that, Taylor & Francis, and particularly Ted Crane, must be congratulated.

It is not coincidental that the first volume in this series is a readings book in transfer of technology, or "international communication of technology," as Professor Robinson aptly calls it. "Technology" has long been recognized as one of the few available solutions, or at least alleviating factors, to combat one of the lingering economic problems of our time—the widening gap between the industrialized countries of the north, and the developing countries of the south.

Even though the recent democratic revolutions in Eastern Europe, by eliminating the "Second World," have somewhat blurred this distinction, the basic premise of the north-south inter-relationships, that is, the income disparity between the two groups, remains unchanged. Thus the international transfer of technology was selected as the theme of the first volume in this series. I look forward to the publication of this book as well as many others of similar importance.

Khosrow Fatemi
Laredo State University
Series Editor-in-Chief

Introduction

The international communication of technology is a vast subject, which a book of readings of reasonable length can cover only very partially. The notion of *communication,* as opposed to the *transfer* of technology, is perhaps a more apt description of what is, in fact, involved. One does not *transfer* technology in the sense of scooping it up in one place and delivering it in another. Unlike a physical good, technology remains where it was spawned. It can only be replicated elsewhere through some sort of communication process. I have defined the process in these words:

> The development by people in one country of the capacity on the part of nationals of another country to use, adopt, replicate, modify, or further expand the knowledge and skills associated either with a different manner of consumption or product use, or a different method of manufacture or performance of either a product or service.[1]

One of the authors whose work is included in this volume has suggested that "it is more accurate to view technology transfer as a relationship, rather than an act."[2] One might say that it is a relationship built by a communication process.

This book owes its inception to Khosrow Fatemi, editor of the *International Trade Journal.* He suggested that the articles appearing in the special Fall 1989 issue of the *Journal* would, with a few additions, constitute a useful addition to the literature and found a publisher. I responded by soliciting friends and acquaintences knowledgeable about one or more aspects of the international communication of technology. Professor Fatemi was kind enough to release the copyrights on the articles that appeared in his journal; they are found in Chapters One, Three, Four, and Nine of this book.

By sheer good fortune, the contributions of the 12 authors whose works are included here flow rather easily from one subject to another—so much so that it might appear that the entire book had been written by a single individual.

1

We start with an exposition of a general model (Chapter One) in which the interrelationship of many of the factors influencing the international communication of technology are made explicit. These variables are related to essentially four corporate decisions: (1) whether or not to communicate technology abroad, (2) if so, the nature of the technology to be so communicated, (3) whether to opt for an internal or external communication, and (4) the selection of the most appropriate mode of communication.

Chapter Two, by William Yager, after discussing various alternative theories or models, empirically tests that model proposed in the previous chapter. He does so by analyzing case studies drawn from four broad industry categories (athletic footwear, food processing, industrial building materials, and electronics). Geographically, he concentrates on East Asia—China, Japan, Hong Kong, and Singapore. In summary, Yager found the model "useful in describing the structure among the profusion of variables influencing technology and transfer mechanism choice."

The contribution by Farok Contractor, Chapter Three, plumbs the depth of some of the strategy choices suggested in Chapter One: whether a technology owner is likely to effect a technology transfer intrafirm (internally) or interfirm (externally). The author looks at a number of factors bearing on that decision, such as the relative ease of appropriating the benefits to be derived from the technology, the risk-adjusted return to be expected from internal as opposed to external transfers, imperfections in the international technology market, transaction costs, company-specific factors (such as the availability of peripheral technology, corporate culture, etc.), pressures in the international business environment (e.g., increasing risk, higher interest rates, accelerating R&D costs, emergence of a technology market), and changes in corporate culture and policy. The latter, it is suggested, is a function of three factors: the ability to manage multicultural enterprises, the ability to detach control from ownership, and the ability to restrain the competitive potential of those enterprises to which technology has been communicated. Contractor speculates on the circumstances under which country policies influencing the choice of communication mode (licensing, joint venture, wholly-owned subsidiary) provide the greatest net benefit to the recipient country.

That discussion leads neatly into Chapter Four, by Jack Behrman, William Fischer, and Dennis Simon. Using the specific example of technology transfer to the People's Republic of China, they establish by case examples the extreme difficulty, which in fact results in higher than necessary costs, of effectively communicating technology in the absence of two key conditions. The first condition is the ability and willingness of the foreign firm to effect a complete transfer, which includes appropriate training, spare parts supply, and follow-through maintenance. The second condition is a willingness, in this case on the part of the Chinese enterprises to which the technology is being communicated, to adopt a learning posture. Here, we are dealing with the relationship between the transfering and recipient firms.

Chapter Five, by Joseph Battat, focuses on the policy, programs, and institutions designed to foster backward linkages between foreign affiliates and domestic firms in four countries—South Korea, Taiwan, Singapore, and Ireland. The backward linkage between a foreign affiliate and local firms refers, of course, to the dissemination of technology after being communicated to the initial recipient. This discussion is based on a survey of backward linkages in the four countries in the fall of 1989 via interviews of government and paragovernmental agencies, samples of foreign affiliates and domestic firms, and business and industry associations. It was found that all programs were closely linked to prior existing small and medium-size enterprise (SME) development programs. In his conclusion, Battat poses a key question: "In the design and implementation of market-oriented SME and linkage programs, how to minimize market distortions so as not to support economically nonviable SMEs and linkages?" We are indebted to the Foreign Investment Advisory Service in the World Bank for having released this study for publication here.

Publically sponsored efforts to broaden and deepen the technology base of Canadian industry are described by Isaiah Litvak in Chapter Six. Although Canadian concern for the promotion of indigenous R&D is long-standing, only recently have Canadian governments stressed the need to encourage trade associations and industry research associations (IRAs) to assume a more active role in the effort. Litvak points out that "a major weakness in Canada's technological policy is the focus on domestic production of new technology while the processes of adoption and diffusion are neglected." He is speaking of the same "backward linkages" analyzed in Chapter Five. The bulk of Litvak's chapter deals with the activity of four publicly assisted nonprofit organizations charged with carrying out and/or facilitating R&D related to the problems of specific industrial sectors. The four are the Canadian Plastics Institute, the Canadian Steel Industry Research Association, the Canadian Gas Research Institute, and the Pulp and Paper Research Institute of Canada. The federal government played a central part during the start-up of each of the four. Litvak points out that such organizations "can play a vital role as knowledge networks for small and medium-sized firms [the SMEs of the Battat chapter] by facilitating technology transfer through the available technical information, advice, and know-how." One problem is that these industry-based, nonprofit R&D organizations tend to be taken over by the larger firms, those with their own R&D, and thus are in a better position to profit.

Michael Bernhart, who has had long experience himself as a technology transfer agent, addresses in Chapter Seven the problem of selection and training of transfer agents. Thus far, we have been talking as though technology were somehow communicated mechanically. In fact, of course, the effective communication of technology internationally requires very special skills, a form of "soft" technology in itself. Bernhart describes an approach that "technology transfer agents" may use to identify behavior relevant to success in the communication of technology and how that behavior is manifested in the society within which one is working. This approach involves "systematic observation, questioning, and experimentation." In that few individuals charged with communicating technology internationally are forewarned far enough in advance to undertake serious training in respect to the relevant culture, most frequently those responsible for communicating technology across cultural boundaries must learn on the job. Many fail. But Bernhart argues that the risk of failure can be reduced by seeing that which is significant, asking the relevant questions of the right people, and occasionally probing the system to test behavior—that is, to check to ascertain what one sees or is told is, in fact, the reality. He suggests a number of key factors to which one should be alert.

Bruce Morgan, in Chapter Eight, develops this subject from a somewhat different point of view when he discusses the transfer of "soft" technologies, which he defines as "learned behaviors: skills and methodologies that do not have fixed technical meanings or formulas, and may even be subject to ambiguity or controversy in their original environment." He cites several examples: "management . . . manpower development and related topics, organizational development and its related disciplines, and many innovative practices in any field, whether it be health, agriculture, or social welfare." This definition should be kept in mind when reading Chapter Ten, which deals with the international communication of technology within the context of corporate decision making. Morgan concludes that "the key element in soft technology transfer is designing a transfer process that anticipates obstacles and successfully adapts the technology to match the receiving environment." He would, I think, agree that these two considerations are more important in the communication of "soft" as opposed to "hard" technologies. The latter may permit some room for error and subsequent adjustment. One may not have that degree of freedom in the former.

A classic example of the international communication or transfer of soft technology is described in Chapter Nine by Eleanor Westney. She focuses specifically on the transfer of organizational structures and processes, what she calls, "organizational technologies," which fits well within Morgan's definition of "soft" technology. Westney points out that with the increasing internationalization of the service industries and the recognition by corporations operating internationally that one of the key elements in their international competitiveness is the way in which they organize tasks and people. Additionally, as the author points out, there is growing interest in emulating the organizational technologies of those firms perceived as successful international competitors, whether they be domestic or foreign. The international transfer of organizational technologies is of special difficulty for several reasons. First, they carry with them the values and behavior patterns of the originating society, which may well be at odds with the values and behavior of the recipient society. Second, the impact of the transfer on the power and position of individuals and groups within the adopting organization can be of signal importance. Westney demonstrates the importance of various factors influencing the transfer of organizational technologies by drawing examples from Meiji Japan, during which era a wide range of foreign organizational forms were transferred to that country. Her conclusion is that these cases "provide a beginning for creating a typology of change factors that shape the adaptation of transferred organizational technologies." She concludes by urging a more precise analysis of the process of such transfer.

The Westney chapter is particularly relevant to the current discussion of what elements of "Japanese management" can be used profitably by non-Japanese firms to improve their international competitiveness. Can certain organizational technologies be transferred effectively into American or European firms? These considerations carry us back to re-examine the model proposed in Chapter One, which refers to the choice of technology to be transferred, both on the supply and demand sides. The model is deficient, we find, in not specifying explicitly that organizational technology is one of the choices and that "environmental specificity" includes cultural "fit" of the transferred technology. Finally, the model should possibly expand the "organizational technology" concept by specifying three distinct types, as does Westney: (1) physical technologies supported by certain organizational technologies (such as telephone systems), (2) organizational technologies supported by certain physical technologies (such as a postal system), and (3) purely organizational technologies (such as quality control circles). Morgan, in his several examples in Chapter Eight, says very much the same thing in somewhat different words.

Chapter Ten, by Richard Robinson, proposes a way of looking at corporate strategic decision making that relates international technology communication to both organizational theory and to the theory of direct foreign investment. It is argued that the process by which decisions are rendered in respect to the location of corporate functions (equated with applied technologies) and to the extent to which technology is traded separately from capital and goods consists essentially of five steps. These steps have to do with identifying the separable links in the firm's value-added chain, determining the source of the firm's true competitive advantages (considering both economies of scale and scope), ascertaining the level of transaction cost between links in the firm's value-added chain (internal and external, domestic and international), determining the comparative advantages of countries relative to the location of each link in the value-added chain, and developing adequate flexibility in corporate decision making and organizational design to permit the firm to respond to changes in both its competitive advantage and in the comparative advantages of countries. Much of the analysis has to do with the communication of "soft" technology, that is, learned organizational behavior, including elements of management, all of which is neatly defined by Westney and Morgan.

Chapter Eleven, by Tagi Sagafi-nejad, is essentially a survey of the relevant literature. The author not only reviews the literature, but also assesses and categorizes it and identifies areas of useful future research. Appended is an extensive bibliography of English-language works. The author notes the major themes in the technology transfer literature of the past decade. These themes include the changing technological balance of power, the role of patents, the regulatory environment, technology transfer from the Third World, and noncommercial transfers. He suggests a conceptual framework considered useful to screening the literature. To that end, four sets of variables are proposed: technology, transfer, organizational, and environmental. Contributions under each heading are discussed. The chapter then moves to consider "cost and compensation issues." Although the literature on international technology movement is exploding, the author nonetheless feels that there is need for further research in respect to technology transfer in the service industry (we are back to the "soft" technology concept once again) and in what he calls "managerial dilemmas." Some of these latter are the trade-offs between various modes of transfer, the pricing of technology, and the terms of technical collaboration agreements.

So, little by little, experience accumulates and theories build. Unfortunately, detailed case studies of international technology communication, firm by firm, are still needed. The underlying negotiations, history of interpersonal relations, and the texts of actual agreements are so sensitive that companies are disinclined to cooperate in analysis of their experience. A personal effort to put together case books on the subject proved to be exceptionally difficult. One had to relay on dated or heavily disguised situations in most instances. Yet one is increasingly aware of the massive communication of technology internationally and the political, economic, legal, and cultural problems generated thereby. Many political observers would credit the spread of contemporary science and technology, and the benefits promised thereby with being the principal force in initiating perestroika and revolt in the Soviet Union and Eastern Europe, which events were hardly anticipated by even the most astute observers.

What is bursting upon the political scene is an awareness that the impact of technology is such that it forces a *global* response to the issues raised, whether they relate to the environment, energy, food, prolongation of life, arms systems, exploitation of the arctic, and deep sea, exploration of the universe, or whatever. Our present perceptions of international technology communication are inadequate for they rarely contain any element of responsibility for the *impact* of what is being communicated. Inevitably, public agencies above and beyond mere national governments must become prime actors in the stimulating, monitoring, and, in some instances, controlling the spawning and dissemination of technology. Unfortunately, this book cannot cover these critically important dimensions of the subject.

Richard D. Robinson
University of Puget Sound
Tacoma, Washington
Volume Editor

Notes to Introduction

1. Richard D. Robinson, *The International Transfer of Technology* (Cambridge, MA: Ballinger Publishing, 1988), p.10.
2. Farok J. Contractor, "The Composition of Licensing Fees and Arrangements as a Function of Economic Development of Recipient Nations," *Journal of International Business Studies*, 11, no. 3, Winter 1980, 47.

Chapter 1

Toward Creating an International Technology Transfer Paradigm

Richard D. Robinson

INTRODUCTION

For anyone studying the literature on international technology transfer, it quickly becomes apparent that what is lacking is an overlying model of the entire process, which, in the final analysis, is a decision-making process at both the supply and demand ends of such transfer.

Clearly, for an international technology transfer to take place, an organization in one country must have technology wanted by an organization in another country. (Here we confine ourselves largely to transfers in which either the supplying organization is a profit-maximizing entity, or the organization on the demand end is such, or both. Excluded from most of the discussion are intergovernmental transfers effected essentially for political purposes, or those made by nonprofit organizations.)

It is clear that some corporations have a higher propensity to transfer technology internationally than others, just as some are more inclined to direct foreign investment, of which technology is almost always an important part. Likewise, some corporations are more inclined to transfer that technology *externally,* as opposed to a purely *internal* transfer of the sort implicit in direct foreign investment. An external transfer is one in which the recipient has no long-standing, formal relationship with the transferring firm. That is, it is a market-mediated transaction. But even if the firm opts to make a transfer, a decision remains as to precisely what technology is to be provided. Similar differences among organizations can be perceived on the demand side. Some are very prone to seek out foreign technology, but even so the transfer may be effected either internally within the organization's own network or purchased from other, unrelated entities. And, as on the supply side, there is the further question of precisely what type of technology

9

is desired. Finally, linking the supply and demand sides in the international transfer of technology are a variety of legal mechanism and organizational intermediaries for effecting the transfer. Where choices exist, managerial decisions are obviously called for. To summarize, the following are among such decisions:

On the Supply Side	On the Demand Side
Decision to transfer technology	Decision to ingest technology
Decision to transfer externally (or internally)	Decision to transfer externally (or internally)
Choice of technology to transfer	Choice of technology to seek
Choice of mechanism for the transfer	Choice of mechanism for the transfer
Choice of organizational link	Choice of organizational link

The relevant questions are: (1) in actual practice what factors tend to determine the choices? (2) What other factors might an organization consider?

DEFINITIONS

Before proceeding further into modeling the process or building a paradigm, one must define more precisely what is meant by international technology transfer:

> The development by people in one country of the capacity on the part of nationals of another country to use, adopt, replicate, modify, or further expand the knowledge and skills associated either with a different manner of consumption or product use, or a different method of manufacture or production of either a product or service, than that previously known.

The definition submitted here admittedly begs many question, for technology has many dimensions. When all is said and done, perhaps "it is more accurate to view technology transfer as relationship, rather than an act."[1] But one might well ask, how does one measure a relationship? Possibly only in terms of the expectations of partners involved. Such speculation leads us to try to define at least the most important characteristics or dimensions of technology; for these characteristics could bear heavily on the nature of the transfer process and on the relationships that might be involved. After surveying much of the relevant literature and talking with a number of knowledgeable people over time, 13 dimensions are presented here, along which technology might be meaningfully described (see Fig. 1.1). Two others,

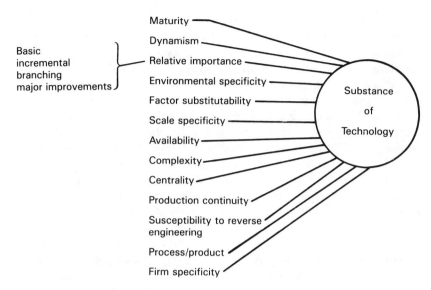

Figure 1.1. Technological dimensions

which are more accurately decision options, are discussed. It is useful to run through these 15 dimensions very briefly, although some are self-explanatory.

The first characteristic, *maturity,* relates to where a technology lies in its life cycle. In this respect, one can identify perhaps 11 phases in the *technology life cycle:*

1. Discovery phase
2. Research phase
3. Development phase
4. Commercial phase
5. Market introduction phase
6. Modification (or adjustment) phase
7. Standardization phase (which may be market specific)
8. Universalization phase (that is, when use spreads to virtually all markets)
9. Competitive intensification phase
10. Replacement phase
11. Disappearance phase

For some technologies, some of these phases may be collapsed in time so as to be virtually simultaneous. This thought leads to the second characteristic, *dynamism,* which has to do with the rapidity with which a technology is changing, or is expected to change—that is, how rapidly it moves through the 11 phases in the technology life cycle.

A third characteristic is *relative importance.* For example, is the technology basic to an entire new development? Or is it of an incremental nature, perhaps representing only a broadening of application via relatively minor adaptation? Or perhaps it is derivative or branching technology, which represents relatively important adaptations. Finally, there is *major* improvement technology. An example of the differences: transition from the invention of manned flight (basic technology), to larger and faster and safer aircraft (incremental technology), to such specialized aircraft as helicopters (derivative or branching technology), to a major jump as represented by jet propulsion, and now rocket propulsion (improvement technology). All are built on the same basic aerodynamic principles, albeit greatly refined.

Such characteristics as *environmental specificity, factor substitutability,* and *scale specificity* are largely self-explanatory. A technology requiring a certain environment (physical, economic, or industrial) to be of commercial value is an example of the first. A new automotive technology is of value only where land vehicles can be used, where there is a perceived need and capacity to buy, and where an adequate industrial and service infrastructure exists. Factor substitutability has to do with the possibility of substituting labor for capital in the use or application of a technology. Scale specificity has to do with the need to produce a given amount of something in order to achieve an acceptable per unit cost of output. One should bear in mind, however, that the apparent limitations imposed by environment on factor substitutability (that is, the production function) and scale may really be the product of the state of technology, not the state of nature.

This discussion brings us to three additional characteristics—the *availability* of a technology, its *complexity* (as measured by the skill levels required to use and transfer it), and its *centrality* (as measured by the percentage of the transferring firm's total earnings related to a given technology). If a technology demanded overseas is readily available, or is at least on the drawing boards, that is one thing. But if it virtually has to be discovered, that is something else, and both cost and risk are very likely to be perceived as higher than in the first case.

A further characteristic, *production continuity,* has to do with the divisibility of a technology. Possibly one technology cannot be employed physically separated from another. One has in mind types of "flow-through" or "continuous process" technology, as in the chemical industry or perhaps various metal extraction, smelting, and fabrication industries. Some technologies obviously have more *susceptibility to reverse engineering* than others. Process technology generally is inclined to be less vulnerable in this respect than product technology, but not invariably so. By "reverse engineering," one is, of course, referring to the relative ease of reproducing a technology from analysis of the product or what is known about it in the public domain. Consequently, a closely related, but distinctive, characteristic of technology relates to whether it is basically *product* or *process* technology. The design, configuration, and physical characteristics of many products may be readily copied by skilled persons. In contrast, some products may contain "black boxes," the reproduction of which may require special materials and skills unique to a very few individuals or firms. Process technology may be relatively more difficult (that is, more costly) to transfer effectively than a purely product technology. More skill transfer may be involved, which can be exceedingly tedious, time-consuming, and requiring special communication and training capabilities. Finally, there is the technology that can be called *firm-specific,* technology known only to a specific firm and that is a product of long years of incremental learning by its employees. They simply know how to do something better than those in any other firm. Very frequently, it may be protected by maintaining secrecy, rather than holding a patent.

It is postulated here that all of these characteristics may be related to a firm's willingness to transfer technology internationally and to its choice of mode, external or internal. An internal transfer, keep in mind, is one taking place between two closely related entities without benefit of market intermediary, within the "hierarchy," to use the term employed by some corporate strategists. An external transfer is just the opposite, one taking place between two independent and otherwise unrelated entities via a market. For example, a firm may readily transfer a bit of mature technology internationally, even externally, if it feels that by so doing it can prolong the life cycle of that technology (at least, in some markets) and squeeze yet more return from its investment in developing it. If it is a less mature technology, the firm may refrain from transfer or, if it does transfer the technology, to do so internally. Or, if it perceives that a technology is highly dynamic, that new technology is on the horizon, the firm may be more willing to transfer that technology to another entity, that is, externally.

Closely related to these 13 technology characteristics are a pair of transfer *options* that spring from two other dimensions of technology—its primacy and completeness. The first of these, primacy, is defined on a continuum running from *user technology* (skill associated with using the technology), to *product adaptive technology* (the skill to adapt a technology to specific conditions of use), to *manufacturing technology* (the skill required to replicate the technology), to *design modification* (the skill required to modify the design of product and/or process), to *design technology* (the skill required to design new products and/or processes, that is, to invent). Lying someplace between product adaptive technology and manufacturing technology is *assembly;* between manufacturing technology and design modification, *manufacture.* These concepts likewise lie along a continuum from the assembly of components, to the assembly of knocked-down kits, to the gradual increase of local value-added toward 100 percent local fabrication. At each stage, the transfer of technology may be complete or partial. One can hypothesize that a technology may be complete or partial. One can hypothesize that a technology transfer is likely to be fairly complete when only user technology is involved, because such would stimulate local demand for the relevant product. As one moves toward the other end—that is, toward the transfer of design modification capability and design innovation capability—the transferring firm might be inclined to transfer less of the relevant technology. The transfer may become only partial as the transferring firm tries to maintain certain leverage vis-a-vis the client firm in order to assure itself that at the latter will perform as agreed, such as (1) remitting royalties and/or fees on a continuing basis even after the last bit of technology is transferred, (2) avoiding competition in the transferring firm's home market or in other markets stipulated as closed to the client firm, (3) not using the transferor's name or brand unless specifically permitted, and, (4) maintaining a certain quality of production. By retaining some technology (in the form of components, undisclosed "black boxes," or certain processes), the supplying firm may be able to threaten withholding that technology unless the recipient performs in a satisfactory manner.

An example might be the transfer of agricultural tractor-related technology. The transfer of *user* technology would probably be fairly complete in order to stimulate demand and encourage the proper use of the tractor. The transfer of the skills to adapt the tractor (and related equipment) to the specific farm conditions within the target market may likewise be important to maximize sales. As demand rises, the appeal of developing a local source is likely to become increasingly attractive, starting with the assembly of components and eventually leading to something very close to complete local manufacture. By then there may be a tendency to hold back the transfer of technology relating to the manufacture of certain parts, and there may be good economic rationale for doing so. But the withholding process provides continuing leverage on the part of the supplying firm. As design and innovative capacity is developed by the client firm, the temptation to adopt a partial technology transfer strategy may become *very* strong, indeed virtually compelling. How else to contain a potential competitor from entering the supplying firm's key markets?

Although these two factors—primacy and completeness—are, on one hand, simply two more technology characteristics, on the other hand, they are different from the 13 previously discussed in that the transferring firm has (or could have) the option as to what it will transfer and how far it will go toward effecting a complete transfer. In the case of the other 13, the transferring firm has little choice, given the state of known technology, other than the decision as to whether to transfer any technology at all.

These brief comments in no way exhaust this aspect of the subject and suggest many areas of possibly rich research toward developing ways of more precisely measuring these characteristics and relating them to company practice, possibly even to return on the assets used in transfers.

STRUCTURE OF SUPPLY AND DEMAND

In any event, having defined the substance of the transfer on which this model was to focus, the next task is to map, in a very general way, those elements encouraging or hindering international technology transfer, first on the supply side, then on the demand side. At this point, the model is a very simple device (see Fig. 1.2). One side is the supply of technology; on the other, the demand. Linking the two are intermediaries and/or linking mechanisms (see Fig. 1.3).

The next step is to explicate on each side the factors relating to the propensity of firms to supply and to seek technology internationally, their propensity to engage in external or internal transfers, and their inclination to select a particular technology either to transmit or acquire. On each side— the supply and demand ends of the transfer—the supplying and receiving entities are presumably influenced by perceived cost, perceived risk, and anticipated benefit. In addition, both parties are undoubtedly influenced by parent and host government policies on one side and by local and foreign government policies on the other. It also seems reasonable at this point to plug in the perceived cost of modifying the technology by either or both parties so as to make the technology more "appropriate," to make the transfer more effective and less costly, and to facilitate assimilation and integration by the user. Considering only the supply side for the moment, a diagrammatic representation is expanded (see Fig. 1.4). The demand side is simply the flip side of the same set of relationships, although with some minor differences. These relationships are depicted in Figure 1.5.

RELATIONSHIPS

In examining Figures 1.6 and 1.7, which simply expand Figures 1.4 and 1.5, one can begin to see the complexity of the relationships in the technology transfer process. Obviously, in a brief presentation, one cannot discuss each one, but it should be pointed out that each relationship suggested

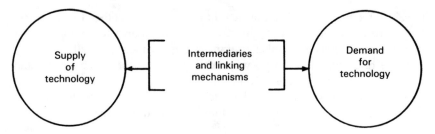

Figure 1.2. Structure of the initial, simplified model.

here could be the subject of valuable empirical research. Only a few have really been examined in detail. A useful way to read one of these exhibits, to take Figure 1.6 as an example, is to say that a firm's propensity to transfer technology internationally is a function of parent government intervention, company history (that is, past experience), company resources, availability of external finance, perceived cost, perceived risk, and anticipated benefits,

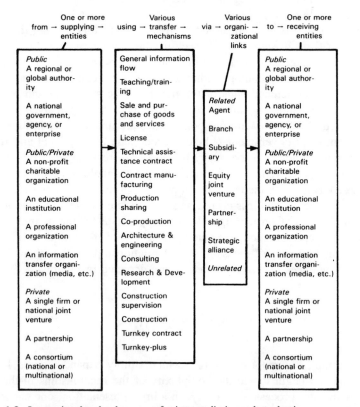

Figure 1.3. International technology transfer intermediaries and mechanisms.

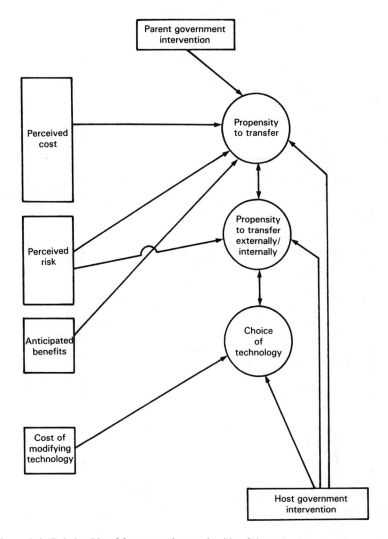

Figure 1.4. Relationship of factors on the supply side of the technology transfer process.

the firm's awareness of foreign demand for technology, host government intervention, plus an *interactive* factor, the firm's propensity to transfer externally or internally. (The fact that it has in place an internal channel through which to transfer technology abroad may influence its decision to transfer technology in the first place.) In turn, perceived cost is hypothesized to be a function of the firm's past relationship with the proposed recipient, the ownership of the proposed recipient (public or private), the labor/capital intensity of the technology (its complexity), the availability of external fi-

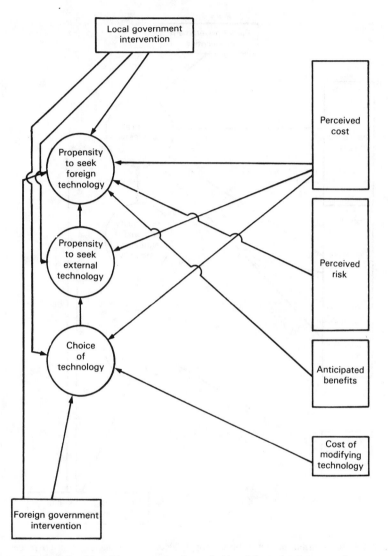

Figure 1.5. Relationship of factors on the demand side of the technology transfer process.

nancing for the transfer, length of experience with the subject technology, the size of the firm, the adequacy of the firm's cost accounting system (to measure the transfer's return on corporate assets committed to the transfer), and the number of prior technology transfers.

As indicated earlier, some empirical research has been compiled over the

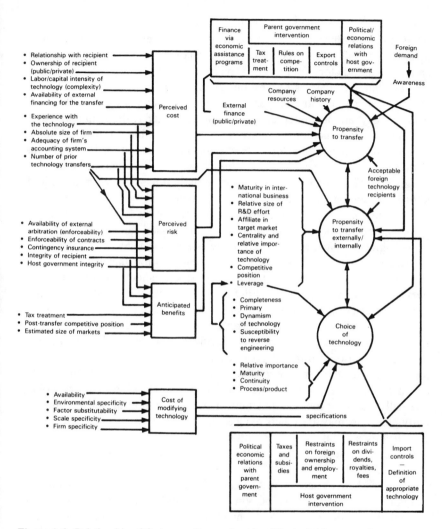

Figure 1.6. Relationship of factors on the supply side of the technology transfer process.

past decade or two relating to some of these relationships. For example, it has been demonstrated that the maturity of a firm, that is, its experience in international markets, may in fact relate to its choice of technology transfer mode—whether it opts for an external transfer or internal. It would seem that during the earlier years of a firm's international involvement, it is inclined to have a higher propensity to transfer technology internationally to largely unrelated entities, that is, externally. As it gains in international maturity, it is more inclined to seek internal transfers, but that tendency drops off as it gains further in international maturity. This relationship can be shown as in Figure 1.8. This type of changing relationship is implicit in the apparently contradictory findings of Davidson and McFetridge and of Zenoff.[2] One can but speculate as to why this relationship between firm maturity in international business and propensity to transfer technology externally should hold. In its years of limited experience internationally, the firm may be under pressure to penetrate as many markets as possible within the shortest feasible time. It has developed few associated firms abroad at this time. Furthermore, it has not developed the capacity to manage and control an integrated international system. Hence, initially it has a strong propensity to transfer technology via license and/or technical collaboration agreement to largely unrelated entities overseas. As it learns how to operate in foreign markets, develops some associated foreign firms, and captures the capability of managing an integrated system, there is a strong tendency to internalize all transfers possible. The idea gains currency within management that in order to exercise adequate control it must exploit its technological resources *internally*. As the firm gains still further experience internationally, management may come to realize (*if* its accounting system is up to the challenge) that it can maximize returns on corporate assets committed to overseas markets by limiting exposure and transferring technology to largely *external* entities via various forms of contractual relationships. In fact, by dealing repeatedly with the same unrelated firms overseas, it may minimize transaction cost and risk to the same extent that it might if moving technology internally. Given problems inherent in political exposure and product liability, the risk may be much reduced in the external transfer case. These conclusions are put forward as little more than working hypotheses; they require more research. More of the relevant research is detailed in Robinson.[3]

The hypothetical nature for virtually all of the relationships diagrammed in Figures 1.6 and 1.7 must be admitted. However, in some cases, as with the policies of the relevant governments, the relationship to a firm's perceptions of cost, risk, and benefit is possibly self evident. For example, a firm may be blocked from exercising a policy option if its parent government refuses to permit the export of the technology for strategic or foreign policy reasons; or if the host government refuses its import on the grounds that it is inappropriate in some way.

INTERMEDIARIES AND LINKING MECHANISMS

Figure 1.3 diagrams the flow of technology from one or more supplying entities using various transfer mechanisms via various organizational links

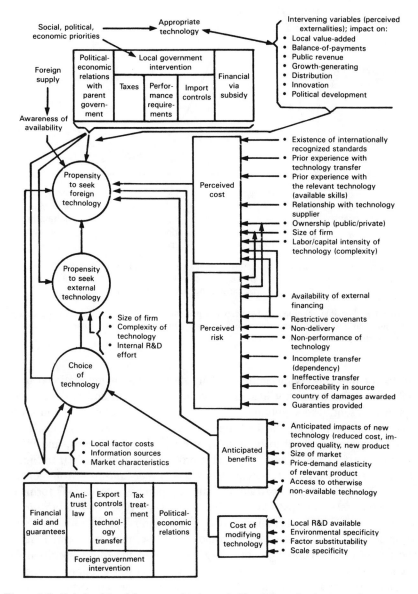

Figure 1.7. Relationship of factors on the demand side of the technology transfer process.

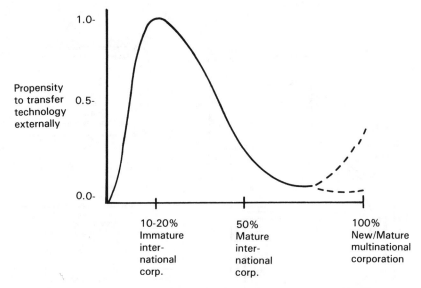

1.0-

Propensity
to transfer   0.5-
technology
externally

0.0-

10-20%	50%	100%
Immature	Mature	New/Mature
inter-	inter-	multinational
national	national	corporation
corp.	corp.	

International experience as measured by the
percentage of total sales manufactured abroad
by controlled subsidiaries.

Figure 1.8. Possible relationship between the propensity of a firm to transfer technology externally and international experience as measured by the percentage of total sales manufactured abroad in controlled subsidiaries.

to one or more receiving entities. Many discussions of international technology transfer treat only inadequately the divers legal and organizational options present. The supplying entity may be in the public sector, in a private sector, or in both. In the latter category fall nonprofit charitable organizations, educational institutions, professional organizations, and information transfer organizations (such as a professional publication). In Figure 1.3 are listed 15 different categories of *transfer mechanisms*—from general information flow, teaching/training, sale and purchase, to various forms of contractual relationships. One can list at least six types or *organizational links*—agent, branch, subsidiary, equity joint venture, partnership, and strategic alliance. An unrelated entity could, or course, take any form. Finally, the nature of the receiving entity may take any of the forms specified for the supplying entity. Although Figure 1.3 implies a one-directional influence emanating from the supplying entity, in fact, there are all sorts of counter influences present. For example, the nature of the receiving entity may influence the nature of the supplying entity. The type of organizational link used in the transfer process may well influence the nature of the supplying entity. Again, the direction and strength of these various influences suggest significant research areas that have been left largely untouched. It is apparent that many legal, managerial, and financial factors are involved in the design of an optimal path for the international flow of a particular technology.

CONCLUSION

By putting together Figures 1.3, 1.6, and 1.7, one has erected a generalized model of international technology flow. Obviously, no single transfer engages all of the factors shown. Many may be completely irrelevant to a given transfer. But the objective here is to develop a *general* model with convincing explanatory power for *all* transfers. As further research is done on the international technology process, the findings can be plugged into the model, either justifying or modifying the relationships suggested, or possibly introducing new ones.

Notes to Chapter One

1. Farok J. Contractor, "The Composition of Licensing Fees and Arrangement as a Function of Economic Development of Technology Recipient Nations," *Journal of International Business Studies,* 11, no. 3, Winter 1980, 47.

2. William H. Davidson, and Donald G. McFetridge, "International Technology Transactions and the Theory of the Firm," *Journal of Industrial Economics,* March 1984, pp. 253–64; reported earlier in William H. Davidson, Amos Tuck School of Business Administration, Dartmouth College, Working Paper No. 106, 1981; and David Zenoff, "Licensing as a Means of Penetrating Foreign Markets," *IDEA,* 14, no. 2, Summer 1970, 297.

3. Richard D. Robinson, *The International Transfer of Technology: Theory, Issues, and Practice* (Cambridge, MA: Ballinger Publishing, 1988). Refer also to Richard D. Robinson, *Cases on International Technology Transfer* (Gig Harbor, WA: Hamlin Publications, 1988).

Chapter 2

An Empirical Study of Variables Related to International Technology Transfer

William F. Yager

A deeper understanding of the process of international technology transfer is at the core of two fundamental and interrelated global relationships: (1) the role of technology in economic development, and (2) the role of transnational corporations as technology suppliers.

As implied by the literature survey in Chapter Eleven, the independent bodies of theory relative to economic development, technology and innovation, and strategic management are all linked together in the study of international technology transfer. An appraisal of the transmission of technological know-how across national and cultural boundaries is made even more intricate by the conceptual overlays of at least three contextual fields of interest: (1) international studies, (2) international trade, and (3) communication theory.

The purpose of this chapter is to describe the evolutionary flow of technology transfer models and to superimpose the Robinson model, described in Chapter One, as a template over a series of recent case studies.

CURRENT RELEVANCE

Probing more thoroughly the mechanisms of transferring technology across borders and the motivations inherent in choosing one approach over another seems even more critical in light of recent dramatic political, social, and economic events in China, the Soviet Union, and Eastern Europe. The rapidity with which fundamental and historic change is occurring only serves to intensify the need to better understand the nuances of the technology-development-corporate linkage.

Within developed economies as well, tensions are building as European Common Market nations move toward a more united economic, if not political, front. Japan's mounting trade surplus is a too-frequent reminder of entire industries lost to global competitors due to ineptness in understanding technology transfer at a more sophisticated level.[1]

TRANSNATIONAL CORPORATIONS

Though performed by a number of public and private organizations, international technology transfer is principally carried out by transnational corporations (TNCs) in the course of their global manufacturing and marketing activities. TNCs as suppliers of technology, capital, and management are inexorably bound to developing countries as suppliers of labor, raw materials, and markets. Entire papers have been dedicated to the issue of delineating the transnational corporation. However, four types of definitions seem to emerge: (1) an ownership definition incorporates a threshold of ownership or control of income-generating assets in more than one country, (2) a structural definition addresses multinationality in terms of the organization of the firm, (3) a performance criterion considers some relative or absolute measure of international spread, such as the percentage of revenue or profit derived overseas, and (4) a behavioral criterion seeks to determine the corporation's degree of geocentricity.[2]

As Buckley observed, definitions are not necessarily right or wrong, just more or less useful.[3] Many of the definitional convolutions are transcended simply by defining a transnational corporation as a firm that owns outputs of goods and services originating in more than one country.[4] For the purposes of this work, this definition is the most useful.

TECHNOLOGY CONCEPTS

Boulding characterized technology as the speeding up of the day when everything is gone.[5] Perhaps, then, it is from the perspective of their 4,000-year history that the Chinese included science and technology, in 1978, as one of their Four Modernizations.[6] The drive for development throughout the majority of the global population considered to be less developed has focused attention over the last three decades on the critical role played by technology in the industrialization process. In contrast, increasing awareness of the erosion of U.S. dominance in technology has precipitated a new level of interest in the forces of technological development in the industrialized nations as well.

From his perspective, Schumacher said simply that technology is how something gets done.[7] Baranson carefully circumscribed an "intricate set of detailed knowledge, skills, and specifications of product designs, production and processing techniques, and meaningful systems used to manufacture particular industrial products."[8] Reflecting the viewpoint of the International Labour Organisation, Singer similarly detailed skills, knowledge, and procedures to produce, what he termed, "useful" goods and services.[9]

Technology is frequently described as embodied in products or disembodied as a product in its own right.[10] Kedia and Bhagat distinguish technologies embodied uniquely in a person, a process, or a product.[11]

Brooke conceived three concentric layers of technology packages. At the periphery are found fairly broad functional (finance, marketing, and management) technologies. In the next layer down, he included processes, products, supplier technicians, and documentation, virtually the media of technology. At the innermost level, the heart of the transfer package is composed of actual technology recipient know-how.[12]

Robinson envisioned four components of a technology transfer package, composed of any combination of: (1) indivisible modules of either core (indispensable) or peripheral (all other) technology, (2) permission to use rights or knowledge, particularly under license, (3) hard goods, typically equipment, or other intermediate or final products having embodied technology, and (4) soft goods, such as documents or oral communication, considered to be disembodied technology.[13]

Proprietary technology is exclusive to the firm, protected by patent, copyright, or trademark. Nonproprietary technology is in the public domain.[14] Bundled technology is considered part of the direct foreign investment package, whereas unbundled technology may be sold separately under contract.[15]

For the purposes of this chapter, technology is considered to be a manufacturing process, encompassing the total operation to generate a product by an overseas affiliate. For the technology to be considered transferred, the manufacturing facility must be fully functioning and indigenous workers adequately trained.

TECHNOLOGY TRANSFER MODELS

As the study of international technology transfer has evolved and matured over the years, models developed to represent the process have become progressively more comprehensive. They have also incorporated a broader recognition of noneconomic influences, political and social factors, and their reciprocal interaction with the transfer decision process.

As an early writer in the field of economic development, Schumpeter in the 1930s envisioned technological change taking place in three stages: invention, innovation, and diffusion.[16] Relating technological change and communication theory a generation later, Marquis and Allen described patterns of technology-associated communication flowing along a time line, also at three levels: science, as a body of knowledge; technology, as a state of the art of applied knowledge; utilization of knowledge, driven by practical need.[17] As shown in Figure 2.1, a series of primarily vertical flows describe the cross-pollenization of knowing among the three levels. Gruber and Marquis later added horizontal flows within each of the three levels.[18] The communication flow labeled "f" is perhaps most representative of international technology transfer,[19] because, over time, it describes the dissemination of knowledge at the practical level motivated by the need for economic development.

Four stages of technological growth were later outlined by Gruber and Marquis to depict the vertical movement from the current level to a new enhanced level of technical knowledge.[20] As indicated in Figure 2.2, the foundational Stage I current technical knowledge base feeds into all successive stages. Invention and discovery at Stage II enlarge the pool of scientific knowledge, forming a new Stage I, as well as cascading into subsequent stages of development. Stage III innovation makes possible the economically driven application of technology to satisfy demand, followed by the diffusion of technological application throughout a society at Stage IV.

In an elaboration of their previous model, Gruber and Marquis subsequently focused on the sequence of steps taken by a firm to move from demand recognition to economic use of new technological information.[21] Figure 2.3 illustrates the process leading up to the action point of tapping into the current pool of knowledge, and then, through adaptation and diffusion, adding back to it. As later described by Robinson,[22] in more elaborate detail, the generalized model of Gruber and Marquis can be adapted to the international context by considering the technology recipient's search process or by highlighting the technology supplier's environmental scanning for an acceptable overseas affiliate. The provision and absorption of new technology by supplier and recipient, respectively, are analogous to the search for and use of the current state of technical knowledge, with two firms instead of one.

Using a three-tiered approach similar to Marquis and Allen, Rogers developed a communications flow model incorporating levels of Basic Research, Applied Research, and Practice (Fig. 2.4), but with international transmission moving specifically from a developed nation to a less devela

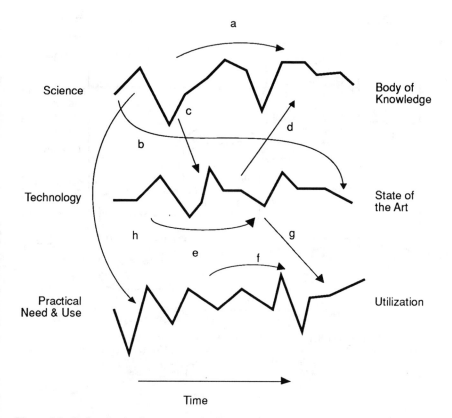

Science

Body of
Knowledge

c

d

b

Technology

State of
the Art

h

g

e

f

Practical
Need & Use

Utilization

Time

Figure 2.1. Paths of technology communication.
Source: Adapted from D. G. Marquis and T. J. Allen, "Communication Patterns in Technology Transfer," *American Psychologist,* 60 (1966), 1053. © 1966 by the American Psychological Association. Adapted by permission.

oped nation, rather than evolving over time.[23] Cross-national communication of research flows horizontally and typifies the content of most international technology transfer efforts.

Placing technology transfer in the context of economic growth, Solo emphasized the universality of transferring "superior" technologies as a prime determinant of development.[24] He described a kind of technological communication cascading down from the realm of pure science to fruition in economic growth (Fig. 2.5). Solo emphasized the critical role played by communication at each step of technological evolution, as information is transmitted and received at each interface. Consequently, technology transfer, placed in a communications context, is not a problem unique to the developing world, but rather a problem commonly shared by suppliers and recipients having differing frames of reference.

Rogers' Research Utilization Process (Fig. 2.6) can be adapted to the in-

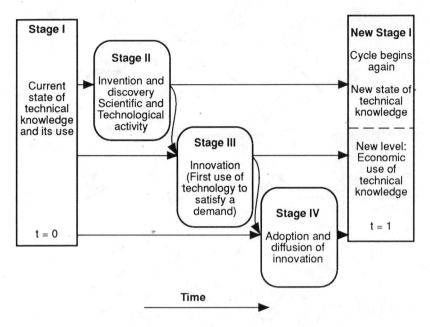

Figure 2.2. The Four Stages of Technological Growth
Source: Adapted from W. H. Gruber and D. G. Marquis, "Research on the Human Factor in the Transfer of Technology. In W. H. Gruber and D. G. Marquis, ed., *Factors in the Transfer of Technology* (Cambridge, MA: MIT Press, 1969). © MIT Press. Adapted by permission.

ternational technology transfer context by considering the Research System to be the R&D function within a transnational corporation and the Client System to represent technology demand of an overseas affiliate or host government. The intermediate Change Agent role is frequently fulfilled by regional (e.g., Asia-Pacific) general managers as they provide an interface between corporate strategies and affiliate implementation. Engineers and technicians also perform in this capacity as their assignments are rotated between affiliate and headquarters locations, exchanging not only technical information, but also cross-cultural perspectives.

In a related vein, Brasseur's study of communications barriers to the transfer of technology focused not only on language differences, but also on communication methods and differences in professional qualifications between suppliers and recipients.[25] As seen in Figure 2.7, the transmitter with superior technical expertise typically attempts to communicate at the top of his or her level of knowledge and may miss the receiver's level of comprehension entirely (message 1). When it is evident that no communication is taking place, the transmitter must lower the level of knowledge sent until the peak of the receiver's level is reached and message 2 is received.

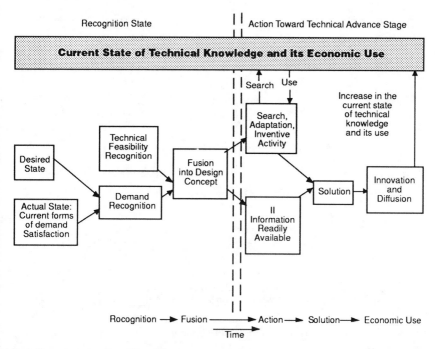

Figure 2.3. Technology development-demand recognition.
Source: Adapted from W. H. Gruber and D. G. Marquis "Research on the Human Factor in the Transfer of Technology. In W. H. Gruber and D. G. Marquis, ed., *Factors in the Transfer of Technology* (Cambridge, MA: MIT Press, 1969). © MIT Press. Adapted by permission.

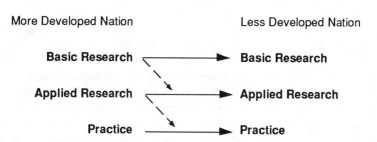

Figure 2.4. The communication of technology.
Source: Adapted from E. M. Rogers, "Key Concepts and models." In K. A. Solo and E. M. Rogers, eds., *Inducing Technological Change for Economic Growth and Development* (East Lansing: Michigan State University Press, 1972).

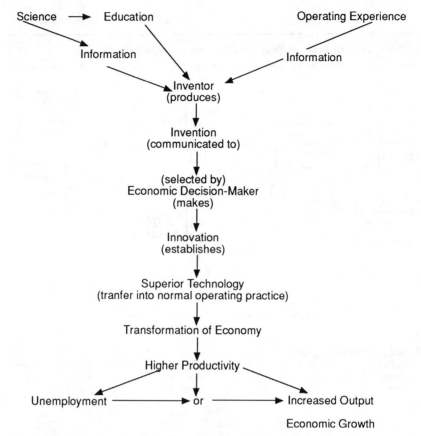

Figure 2.5. Technology transfer and economic growth.
Source: Adapted from R. A. Solo, "Technology Transfer: A universal Process." In R. A. Solo and E. M. Rogers, *Inducing Technological Change for Economic Growth and Development* (East Lansing: Michigan State University Press, 1972).

Unfortunately, in many technological environments, such progressive simplification may not be possible, either because the transmitter is incapable of packaging his or her information in simpler terms or because the sophistication of the subject itself obviates any further reduction (Fig. 2.8). In either case, communication fails, the respective parties are left frustrated, perhaps hostile, and the technology remains lodged with the potential supplier.

The first two cases assume a common culture and language. Even when a technology recipient is able to speak the supplier's language with a degree

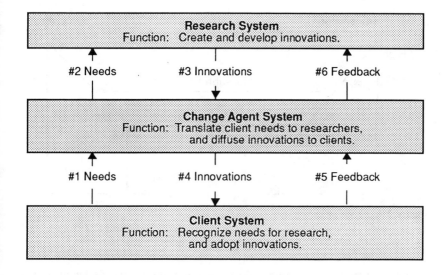

The communication flows numbered in this paradigm are:

#1 Flow of client needs (for information) to change agents.
#2 After interpretation and clarification, these needs are transferred to the research system.
#3 Researchers attempt to provide needed information for clients' needs, either from accumulated knowledge or via newly-originated research.
#4 Change agents distill and interpret this new information (innovations) for clients.
#5 Feedback from clients to change agents on the adequacy of the new information in meeting their needs.
#6 Change agents convey client needs and recycling of the entire process.

Figure 2.6. Paradigm of the research utilization process.
Source: Adapted from E. M. Rogers, "Key Concepts and Models." In R. A. Solo and E. M. Rogers, *Inducing Technological Change for Economic Growth and Development* (East Lansing: Michigan State University Press, 1972).

of facility, cultural barriers may prevent open recognition of a failure to communicate completely, particularly at a detailed technical level. Cultural mores, for example, may block the admission of ignorance and perceived loss of face. Consequently, whether or not an intellectual connection has been made may not be known by the sender in time to retransmit, and essential know-how may be lost indefinitely, yielding a delayed or fatally flawed project.

In a dual-language environment (Fig. 2.9), the receiver has sufficient technical knowledge to understand the transmitter's message if it were in his or her own language or if each party were fluent in the other's language.

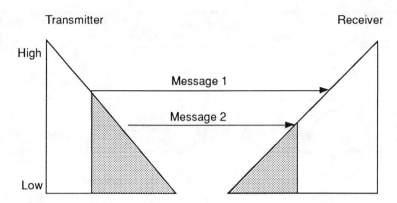

Figure 2.7. Levels of knowledge.
Source: Adapted from R. E. Brasseur, "Constraints in Transfer of Knowledge," *Development,* 18, 3 (1976). 16–18. Adapted by permission of the Society for International Development.

However, when the transmitter does not know the receiver's language and the receiver's knowledge of the transmitter's language is insufficient to grasp the lowest level of the transmitter's message, communication, again, fails.

In contrast, when the transmitter has learned the receiver's language, the receiver's peak level of technical knowledge may not be tapped, but in functioning at the top of his or her foreign language potential, the transmitter has succeeded in communicating message 1 (Fig. 2.10). Cross-cultural obstacles may also need to be overcome in terms of the superior-subordinate positions inferred from a teacher-student relationship. In addition, languages lacking tense and specific technical terms constitute barriers as well.

In the class of procedural models, Bar-Zakey developed a PERT-like net-

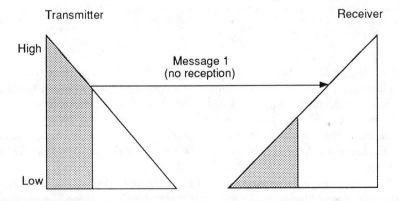

Figure 2.8. Levels of knowledge.
Source: Adapted from R. E. Brasseur, "Constraints in Transfer of Knowledge," *Development,* 18, 3 (1976). 16–18. Adapted by permission of the Society for International Development.

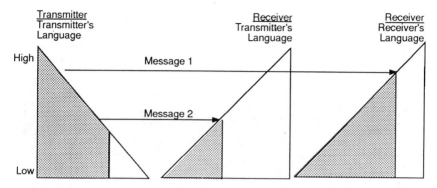

Figure 2.9. Levels of knowledge.
Source: Adapted from R. E. Brasseur, "Constraints in Transfer of Knowledge," *Development*, 18, 3 (1976). 16–18. Adapted by permission of the Society for International Development.

work representing simultaneous activities and goals for technology supplier and recipient.[26] Several unique features of the Bar-Zakey model contribute to its descriptive power, and unavoidably to its complexity (Fig. 2.11). The symbiotic partnership between technology supplier and recipient is recognized as the project progresses. A succession of stages evolve as a function both of time passing and cost increasing. In the Search stage, both supplier and recipient move independently through a series of steps to define their respective technology capability and need, and to establish a "viable contact." A joint go/no-go decision at the conclusion of each stage either ushers in the successive stage or terminates the process.

The second stage, Adaptation, describes the feasibility/desirability evaluation activities of the supplier and recipient, respectively. As a result of

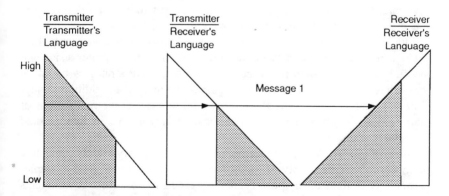

Figure 2.10. Levels of knowledge.
Source: Adapted from R. E. Brasseur, "Constraints in Transfer of Knowledge," *Development*, 18, 3 (1976). 16–18. Adapted by permission of the Society for International Development.

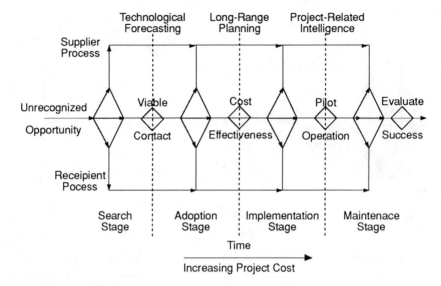

Figure 2.11. Technology transfer model.
Source: Adapted from S. N. Bar-Zakey, "Technology Transfer Model," *Technological Forecasting and Social Change,* 2, 3/4 (1971). © 1971 by Elsevier Science Publishing Co., Inc. Adapted by permission.

positive cost effectiveness decisions by both parties, their efforts then move to the third stage, Implementation of the new technology, culminating in a pilot operation. A successful pilot opens up the fourth and final stage of Maintenance, as the transferred process shifts into full power.

Bar-Zakey, although necessarily oversimplifying the process to fit the linear nature of the PERT mechanism, nevertheless succeeded in capturing a new threshold of more realistic complexity, particularly in the dual perspectives of technology supplier and recipient.

Although not a graphical model, Baranson's analytical framework (Fig. 2.12) acknowledged a variety of constraints, both on the supplier's side and the recipient's side, then proposed a series of relevant strategies based on the parties' objectives and relative bargaining strength.[27] In addition to supplier and recipient firms, the Baranson model introduced the motivations of their respective governments as additional driving forces in international technology transfer.

The host government's perspective can, of course, vary widely. One area of concern is for its supply of foreign exchange. China is a good example, particularly in light of the post-Tiananmen credit freeze. Host governments also erect policy structures to support their development strategy and build in incentives, accordingly. Control of the technology is also an issue when host governments intervene to strengthen the bargaining position of the re-

Supplier Enterprise (Government)	Technology	Purchaser Enterprise (Government)
*U.S. Government policies: Economic consequences (Erosion of production Jobs) Political-strategic considerations (Transfer of industrial capabilities in strategic sectors)	*Distinctive charcteristics: Quantum and complexity License to manufacture or turnkey-plus Operative-duplicative- innovative General-firm-system specific Stage in product/process cycle	* Government policies: Industrially advanced nations Socialist countires Developing nations
*Bargaining power (enterprise): Financial resources and international experience Technological lead and world market position of enterprise		*Bargaining power: Absorptive capabilities Alternative sources of technology Astuteness Financial resources
*Enterprise strategies: Shift from equity invest- ment and management control of sale of technology and manage- ment services Necessity to accept for- eign affiliate due to enormity of R and D of capital investment costs, offset require- ments, or because scale of operation requires consortium Measured release of core technology Sales of technology no longer central to company business or commercially advantageous		*Enterprise strategies: Internationally competitive technology Duplicative and/or innovative design and engineering capabilities Entry into export markets Training of technical managerial manpower Fast, efficient technology transplants

Figure 2.12. Analytical framework: Technology transfer by U.S. corporations.
Source: Baranson, *Technology and the multinationals* (Lexington, MA: Lexington Books, 1978).

cipient firm. Baranson's model also incorporates the host's drive for rapid technological development and notes the emphasis on developing an indigenous engineering capability, as well as acquiring foreign capital investment. Singapore's strategy to develop an independent technological base, described below, is an example of this aspect of Baranson's model.

Supplier-side governments, conversely, are concerned about the loss of investment and employment potential and the possible negative impact on balance of payments. In addition, home country governments may try to restrict the outflow of certain technologies on the basis of political or national security considerations.

In considering the perspective of the technology supplier, the Baranson model builds a strategic context within which the transferring firm assesses risk (e.g., the probability of creating a new competitor), technology characteristics (e.g., maturity and firm specificity), and the transfer mechanism (e.g., licensing a mature technology but controlling a proprietary technology with an internal transfer channel).

Burgleman and Maidique explored the development, application, and degradation of technology in a variety of industries.[28] For a major application in the industry as a whole, the technology life cycle (TLC) amounts to the product life cycle for a generic product as it moves through time-related stages (Fig. 2.13). One of the novel features of the TLC model is its consideration of technology disembodied from products, perhaps even conceived as an independent profit center. This distinction is particularly relevant as the cost-risk-benefit-related propensity to transfer technology is considered in the global environment.

The concept of technology choice driving the transfer process was further explored by Harvey, incorporating the notion of a technology life cycle.[29] The three stages of internal, competitor, and recipient country analysis in Harvey's technology transfer flow model (Fig. 2.14) are somewhat analogous to the pattern in strategic management of assessing the firm, competitive, and industry environment. Harvey conceptualized the level of involvement of the supplier with the technology as dependent on the technology's stage in the life cycle, the terms of transfer, and the internal assessment of the impact of the transfer on operations. Level of involvement with the technology in international transfers implies the degree of intensity in the relationship with the recipient, particularly in terms of transmitting technology software, including the training and development of indigenous workers, perhaps over an extended period of time. Mature technology, later in the life cycle, would be more prone to licensing and external transfer. Newer innovation, perhaps proprietary, or closer to the TNCs technological core, would more likely be transferred internally to maintain control. The impact on the corporation is interdependent with the technology life cycle stage and the feasible terms of transfer, as both engender perceptions of cost, risk, and benefit.

Competitive analysis at domestic and international levels flows into host country appraisal and the analysis of the potential transfer both from supplier and recipient perspectives. Particularly in industries in which global strategies are mandatory to maintain competitive advantage, these processes are probably intensely interrelated, rather than unidirectional, as depicted in the

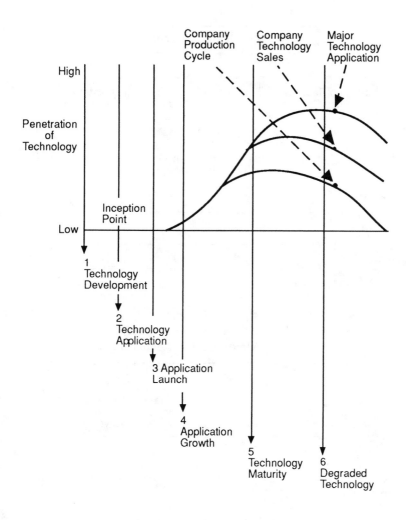

Figure 2.13. The technology life cycle.
Source: Adapted from R. A. Burgleman and M. A. Maidique, *Strategic Management of Technology and Innovation* (Homewood, IL: Irwin, 1988).

model. As an example, the choice to re-enter China following the crackdown of June 1989, has been described by some firms as a response to Japanese attempts to expand their market presence in the post-Tiananmen vacuum. Without the threat of competitive incursion, the U.S. firms would have preferred to wait longer for more tangible evidence of stability from Chinese authorities.

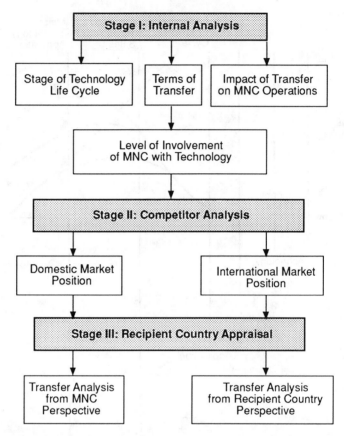

Figure 2.14. International technology flow.
Source: Adapted from M. D. Harvey, "Application of Technology Life Cycles to Technology Transfers." *Journal of Business Strategy,* 5,2 (1984). © by Warren, Gorham & Lamont. Adapted by permission.

Although their study of diffusion research is oriented to marketing applications, focusing on the time for adoption of innovation, Robertson and Gatignon further enriched supply side and contextual factors and the complexity of their interrelationship in the transfer process (Fig. 2.15).[30]

Pennings also described the interaction of innovation attributes and organizational characteristics in terms of their influence on the degree of "radicalness" of the technology.[31] Many of the same variables during this period seemed to emerge in transfer models as well as diffusion models, apparently as part of a growing recognition of the complexity and interactive nature of the process as the field matured (Fig. 2.16).

As procedural, unidirectional models continued to appear in the literature, empirical testing of the myriad of transactions that existed in reality belied

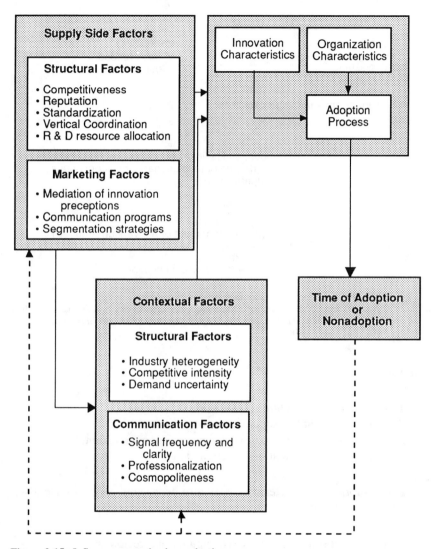

Figure 2.15. Influences on technology adoption.
Source: Adapted from T. S. Robertson and H. Gatignon, "The Diffusion of High Technology Innovations," In J. M. Pennings and A. Buiterdam, eds., *A New Technology as Organizational Innovation* (Cambridge, MA: Ballinger, 1987).

the existence of relatively simple step-by-step transfer processes. In the belief that technology transfers are procedurally unique, Creighton, Jolly, and Buckles attempted to identify common elements in actual transactions.[32] The outcome was the description of nine corresponding characteristics. Effective management of four administrative elements was correlated with successful

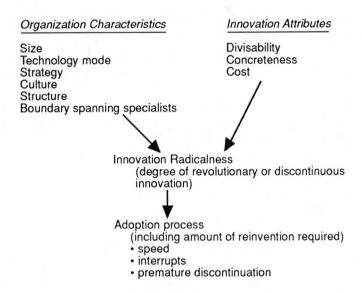

Organization Characteristics   *Innovation Attributes*

Size   Divisability
Technology mode   Concreteness
Strategy   Cost
Culture
Structure
Boundary spanning specialists

Innovation Radicalness
    (degree of revolutionary or discontinuous
    innovation)

Adoption process
    (including amount of reinvention required)
    • speed
    • interrupts
    • premature discontinuation

Figure 2.16. Organization-technology interaction.
Source: Adapted from J. M. Pennings, "Technological Innovations in manufacturing. In J. M. Pennings and A. Buitendam, eds., *A New Technology as Organizational Innovation* (Cambridge, MA: Ballinger, 1978).

transfers:

1. Organization: structure, policies, skills, climate.
2. Project: standards, responsiveness to user needs.
3. Documentation: information, language, communication.
4. Distribution: information system and control.

Five "informal" characteristics were also considered necessary for a successful transfer to occur:

1. Linker: liaison person between supplier and recipient.
2. Capacity: capability, availability of resources.
3. Credibility: perceived integrity of the parties and reliability of information.
4. Willingness: desire, effort to make process work.
5. Reward: perceived mutual net benefit of the transfer.

Samli incorporated many of the elements identified by the Jolly group in his transfer model and described a kind of suspended tension between sender and receiver focused on the technology itself.[33] As senders attempt to bal-

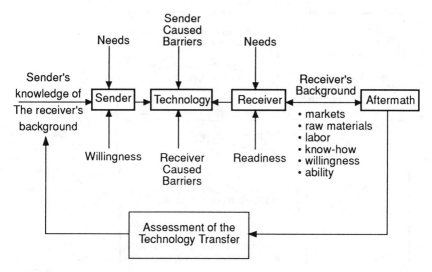

Figure 2.17. International technology transfer model.
Source: Adapted from A. C. Samli, *Technology Transfer* (Westport, CT: Quorum Books 1985). ©
1985 by Greenwood Publishing Group, Inc. Adapted by permission.

ance their needs and willingness to transfer, and as receivers attempt to balance their needs and readiness to absorb the new technology, they both approach the transaction conditioned by their assessment of the impact of prior technology transfers. In other words, there is in Samli's model a recognition of mutual awareness to add realism to the relationship (Fig. 2.17). Samli also added the concept of appropriateness of the technology being considered for transfer by including five broad factors: market characteristics, raw material availability, economies of scale, labor intensity, and machinery adaptability.

Despite the discernable trend toward increased comprehensiveness as technology transfer models evolved, Kedia and Bhagat were virtually unique in their recognition of cultural variations across national boundaries and the associated role played by organizational cultures in determining the effectiveness of technology transfer.[34] Technology dimensions and organizational culture differences were brought together as influences on technology transfer effectiveness (Fig. 2.18). Societal culture differences and the absorptive capacity of the recipient were also described as moderating the transfer process.

In an attempt to synthesize the body of literature on technology transfer by transnational corporations to developing countries, I developed the model in Figure 2.19. In spaning the boundaries of home and host countries, the transnational corporation forms a structural conduit between parent headquarters and overseas subsidiary, and their respective social, economic, and

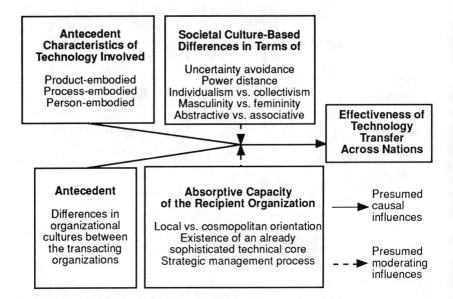

Figure 2.18. Effectiveness of technology transfer.
Source: Adapted from B. L. Kedia and R. S. Bhagat "Cultural Constraints on Transfer of Technology Across Nations," *International Business* (1988).

political contexts.[35] Consistent with the communication analogy, a common field of experience is defined by the overlap of the TNC presence in the host environment, typically composed of indigenous affiliate workers and U.S. expatriates, generally managers and technicians.

Given a propensity to transfer technology of a particular type to a particular affiliate, the TNC headquarters then generates over time a mix of management expertise, worker training, capital, and the technology from a central R&D facility. In return, profits, dividends, royalties, or fees may be received. Overseas, the subsidiary would presumably absorb labor and raw materials to manufacture products for export, earning foreign exchange, or for local or regional markets. Taxes are paid to home and host governments, and political realities are recognized. The local infrastructure is impacted in terms of worker training and new employment levels, as well as new demands placed on transportation, communication, water, and energy systems. The linkage between technology transfer and economic development is also illustrated by growth in the host country.[36]

At the confluence of these efforts, the comprehensive model in Chapter One consolidates many of the foregoing relationships, then moves beyond by incorporating the concepts of propensity and supplier/recipient perceptions in a dynamic interrelationship.

The genesis of the process is the supplier firm's propensity to transfer. A

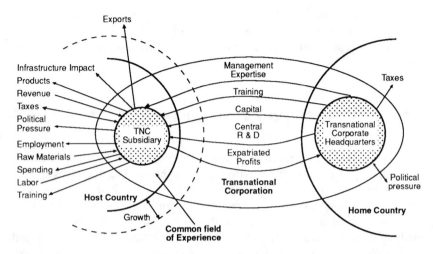

Figure 2.19. International technology transfer.
Source: W. F. Yager, Transnational Corporate Strategy and Economic Development (unpublished).

TNC's inclination to move technological know-how abroad is seen to be a reiterative process conditioned by the choice of the technology to be transferred and the choice of an internal or external mechanism by which the transfer takes place.

The concept of "propensity," both to transfer and to choose a particular transfer mechanism introduced a dynamic and often culturally rooted characteristic more representative of the practioner's international environment.

By including "perception" in the assessment of cost, risk, and benefit, Robinson (Chapter One) incorporated a critical, but nonetheless frequently ignored, quality of cross-cultural transactions, divergent value systems, and varying frames of reference, as well as the potential for dynamic evaluation as new information is absorbed.

COMPARATIVE CASE STUDIES

In order to explore the usefulness of model variables in describing a variety of actual technology transfers, interviews, questionnaires, and archival data were obtained from affiliates in Hong Kong, Singapore, and Guangdong Province in the People's Republic of China, and their headquarters in the United States. Each parent firm had demonstrated a propensity to transfer technology internationally by making direct foreign investments in manufacturing facilities or by entering into contractual relationships with overseas manufacturing affiliates. The same firms had also demonstrated a propensity to withhold technology selectively in each of the three countries based on their assessment of an array of environmental, technological, and internal firm characteristics.

INDUSTRY DOMAIN

Four broad industry categories of athletic footwear, food processing, industrial building materials, and electronics were included on the basis of five criteria: (1) extensive overseas manufacturing activity, (2) diversity in labor intensity, (3) potential for host country economic development, (4) differences in technological maturity and dynamism, and (5) variety of parent-affiliate structural relationships. For example, athletic footwear was selected for its relatively high labor intensity, having experienced overseas low-labor-rate leap-frogging. One manager in a leading firm described one of the key driving forces in the industry as "chasing cheap labor." Another, in Hong Kong, independently characterized the industry as "island hopping" in pursuit of low labor rates. That pursuit has carried manufacturing successively from the United States to Japan, Korea, Taiwan, Indonesia, and China, as each country's standard of living has risen in a series of waves of increasing wages. Now, virtually all athletic footwear is manufactured outside the United States, but to U.S. firm specifications and using U.S. technology, proprietary trademarks, and molds. Overseas apparel and footwear manufacturing is almost exclusively done under contract, and equity joint ventures, common in other industries, are virtually nonexistent.

Food processing has a long history of overseas manufacturing, technology transfers, and cross-cultural relationships. Generally, a labor-intensive industry but with more capital-intensive technology than athletic footwear, food processing is more directly related to internal development than to export. The use of local raw materials and labor has also contributed substantially to the acceptance of TNC foreign investment by developing countries. Although many less-developed countries (LDCs) would also like to be able to earn scarce foreign exchange by shifting more emphasis to exports, as one Asia-Pacific manager commented, "Who wants to buy Chinese soft drinks?"

Food processing technology is generally considered less dynamic than electronics and was found to be transferred frequently via joint ventures having varying degrees of equity participation.

The more labor-intensive assembly and test aspects of electronics manufacturing are extensively exported to Southeast Asian affiliates. Although integrated circuits may not come to mind as pivotal in the economic development strategies of many LDCs, they do earn much needed foreign exchange, contribute substantially to infrastructure development, particularly in education, and confer on the host country a prestigious image of being "high tech." This developmental differentiation strategy is particularly evident in Singapore and intensely desired in China as one of the Four Modernizations.

The development of Asian economies in recent years has also resulted in the emergence of substantial markets for consumer electronic devices. Plants established years ago primarily to take advantage of low labor rates to assemble products for export are experiencing a shift in their raison d'etre. Many are now surrounded by suppliers and industrial customers, together serving growing consumer markets in the same region. As the general manager of one TNC electronics plant said, "All the big guys are right here!"

GEOGRAPHICAL DOMAIN

As Japan recovered in the post-World War II years and grew to a position of global economic preeminence, and as other East Asian economies have developed as newly industrialized countries (NICs), interest in the Pacific Basin has matured from a remote fascination with things Oriental to an economic necessity. Some have dubbed the coming years as the Pacific Century, to acknowledge the shift in the global economic center of gravity.

The enormous developmental potential of China and the opening up of the country to foreign investment over the last 10 years have stimulated a virtual scramble for China connections. By now, however, sufficient experience and data have been absorbed in the international business community to signal that the honeymoon is over, and the events in Tiananmen Square served to precipitate caution and a more broadly held realization that the only generalization that is valid about China is that all generalizations should be suspect.

As the enormously complex Chinese giant continues to experiment with mixed economic development strategies over the next decade, a three-steps-forward-two-steps-back type of progress envelopes the future of Hong Kong as never before. Notwithstanding the impact of the official reabsorption of the British colony back into the Chinese mainland in 1997, Hong Kong is inextricably linked to China economically as well as culturally. A very large amount of Chinese exports move through Hong Kong and provide a thriving, if not essential, re-export trade for the Hong Kong economy. Hong Kong, in contrast, is the reverse conduit to China for severely needed capital. Recently, the American Chamber of Commerce in Hong Kong reported that, due to the scarcity of labor there, nearly as many Chinese workers were employed by Hong Kong enterprises in neighboring Guangdong Province as in Hong Kong itself. Even without the prospect of Hong Kong's reunification with China, their symbiotic relationship suggests the inclusion of both in a study of either one.

To provide a more fertile field for model evaluation, the persuasively orchestrated economic development of Singapore is juxtaposed against the more mature and distinctly laissez-faire atmosphere of Hong Kong. Both are island economies with few natural resources, sharing a common British colonial history, but with widely differing government structures and policies. Though with sizeable Malaysian and Indian minorities, Singapore's population is still 70 percent Chinese, and the business community is dominated by Chinese managers and culture.

The government of Singapore has a clearly articulated strategy for economic development, incorporating a body of policy designed to orient the economy to high technology and to achieve the per capita income level of a developed economy by 2000. The heavily promoted subway is the most modern in the world. Posters and banners exhort students and workers to have pride in Singapore, to be productive, and to excel. As an outgrowth, Singaporean workers are not recognized as the lowest paid but as the most productive in Asia by semiconductor manufacturers in the study. The combination of this productivity and government investment incentives has created a massive high technology industry in the country. As one Singaporean manager commented, "We've moved all the heavy industry—the tire and auto manufacturing—across the border to Malaysia." Government incentives are in place to establish low labor rate feeder plants in other countries or to automate lower skill jobs. Intentions are clear and companies are responsive.

TECHNOLOGY CHOICE

In the context of this framework, variables in the general model were evaluated in questionnaire and interview data from 40 affiliate managers across industry technology and geographic lines.

The choice of technology to transfer is a decision contingent on three classes of variables and the transfer mechanism. First, the diversity in a broad range of exogenous host country political, economic, and social characteristics was distinctly reflected in the three geographic areas studied.

All electronics affiliates were located in Singapore or Hong Kong. Guangzhou, in Guandong Province (PRC), hosted athletic footwear contract manufacturing primarily for regional export and some internal consumption. Industrial glass manufacturing in China was shifted between export and internal markets as the need for foreign exchange became more or less intense. Food processors in China were oriented primarily toward internal consumption. Congruence with development goals was a motivation for transfer, but only when translated into viable economic incentives. Consequently, the granting of tax holidays in Singapore was a tangible incentive for foreign investment in high technology manufacturing facilities. In contrast, joint ventures with foreign partners by Chinese state-owned enterprises were seen by the PRC government as conduits for technological development. Only those foreign partners willing to share technology in exchange for access to domestic markets saw these arrangements as acceptable.

Countries with available technical and managerial skills attracted more advanced technologies, and conversely, countries without these skills attracted more labor-intensive technologies. Of the affiliates indicating that these skills were either not at all available or only somewhat available, all were in China. Hong Kong, and secondarily Singapore, firms had minimal problems acquiring technical and managerial skills.

Technology choice was not seen by this set of firms to be influenced by home country government policy to a substantial degree. The pattern of Sino-U.S. relations before and after Tiananmen Square was closely observed, but competitors' decisions to withdraw from China or to remain were perceived to be more influential than U.S. government policy. Tariffs on the movement of electronic subassemblies and finished products into the United States were pointedly used to illustrate U.S. policy in contrast to the genuine free trade orientation of Singapore and Hong Kong.

A group of technology dimensions in the model influences the cost of modification to meet host specifications, supplier leverage, and technology choice. Availability tends to drive the modification cost down as a broader range of appropriate technologies is offered by alternative suppliers. One manager, however, related the story of a Chinese state enterprise attempting to construct, independently, an instant coffee plant from publicly available data in books and articles. Several years and millions of yuan later, the uncompleted facility was sold to a U.S. joint venture. Following the injection of needed managerial and technological expertise, the building was nearly gutted, some hardware was salvaged, and a new process was begun nearly from scratch. This story serves to reinforce the pivotal importance of a complete software component being present to envelope an effective technology transfer.

Environmental specificity of technology being transferred can be seen in the seemingly prosaic requirement for athletic shoes to remain white during assembly. However, in a very dusty region of China, where open windows were the only source of air circulation during hot summer weather, what was white and what was not white was a discussion rooted perhaps as much in culture as in economic necessity.

In contrast, the controlled internal environment required for a semiconductor manufacturer's clean room is the same in Singapore or Silicon Valley.

Factor substitutability is best illustrated by the exchange of low cost labor for capital-intensive machinery, the genesis for much of the technology transfer to Southeast Asia. In addition, however, once a plant is built overseas, it may represent an exit barrier from the host country and retard the (re)substitution of more technologically advanced, capital-intensive processes in the United States.

Following the initial round of direct investment in response to host government incentives, however, the question of continued cost savings remained, in some cases for over 20 years. Corporate inertia, notwithstanding, once an overseas manufacturing facility had been established, the decision continued to be re-examined in terms of cost trade-offs: increased transportation, rework, and liaison costs in exchange for lower labor costs. As one executive commented, "When an operation can be automated, with equivalent results, and capital can be substituted in the U.S. for overseas labor, we'll do it." Host country infrastructures, however, are much more dynamic than static, particularly when responding to a clearly articulated growth strategy in Singapore. Indigineous technical capability has grown in an environment of economic incentives, improved educational opportunities, and more sophisticated, higher paying jobs. A upward spiraling supply-demand relationship has resulted in the concentration of technical design capability to supplement productive, efficient manufacturing. The question of relocating automated production to the United States to complement design capability has become a moot issue in many firms, because design expertise, closer to customers, is now a reality in most Asian locations in addition to increased automated assembly.

Scale specificity of transferred technology was observed primarily to be an issue for food processors in China. In the introductory phase of the product life cycle, continuous processing was locally redesigned to adapt to smaller but growing demand for products perceived as new to the culture.

Firm specificity, in the context of wholly owned subsidiaries, is perhaps more appropriately seen as embodied in technical personnel from parent design center locations. Effective transfer was generally the result of frequent and prolonged contact with recipient personal. Subsequently, in numerous firms, rotating assignments of technicians and engineers between affiliate and parent locations facilitated the communication of new ideas and deepened mutual understanding at a cultural as well as technical level.

Maturity or age of a given technology represents a key factor in the choice of technology to transfer. Many global firms with ongoing R&D efforts prefer to transfer older, generally more labor-intensive technology abroad to take advantage of lower labor rates. However, low labor rates are not necessarily equivalent to low labor cost per unit, because rework, transportation, supervision, and extensive fringe benefits often erode actual labor productivity.

With a vision of leap-frogging historical development stages, many host countries prefer, or even demand, only the most advanced technology, whether or not appropriate to the current infrastructure.

Completeness of the technology transfer refers to the scope of knowledge conveyed to the recipient at any given stage of the process. In the context of this study, particularly in the case of wholly owned subsidiaries, all transfers were as complete as possible. In one joint venture in China, the parent firm licensed a proprietary step in the manufacturing process and expatriot managers were present to oversee its security.

The degree of primacy transferred constitutes the progressive ability of the recipient to move from knowing how to use the product to successive stages of product adaptation, manufacturing, design or process modification, and, finally, to the capability to design new products or processes. One frequently overlooked aspect of technology transfers is the enduring relationship between supplier and recipient. Economic and social development in the host environment over a period of years may change radically the appropriateness and primacy of technology being transferred.

In the context of a host country's environmental dynamism is the dynamism of the technology itself. As the "half-life" of a technology's commercial viability is reduced, the web of interdependency between parent and affiliate is drawn tighter. Even though the parent-supplier may have increased leverage over recipients as the developer of newer and newer technologies, it is the parent's dependency on off-shore affiliates for efficient manufacturing that is often the equalizer in such relationships. The result is frequently the cross-pollenization of design, manufacturing, and customer requirements, mentioned above.

As the susceptibility to reverse engineering declines, the supplier will have more leverage in the transfer, because the commodity already developed and for sale may be far less expensive that the R&D investment required to replicate it. One semiconductor manager added an interesting nuance in commenting that even though the design of a product might be determined by reverse engineering, the ability subsequently to produce that product at comparable quality levels may be entirely lacking. As the instant coffee plant illustration above indicated, availability of a design does not necessarily lead to replication of the process.

The mechanism chosen to convey the technology to the recipient influences what that technology will be, and vice versa. TNCs transferring proprietary technology perceived to need additional protection from piracy typically use an internal structure, a wholly owned subsidiary. However, confusion between equity and control seems to be abating as TNCs who have gained more international experience move more toward contractual relationships, particularly as limitations on foreign ownership have forced their use and the consequent broadening of the experience base.

Conversely, where the choice of the transfer mechanism is restricted, for whatever reason, only a narrower range of technologies may be perceived to be appropriate for a given channel.

PROPENSITY TO TRANSFER INTERNALLY OR EXTERNALLY

Endogenous firm variables, both supplier and recipient characteristics, play major roles in the choice of transfer mechanism. Home and host government policies are believed to be substantial influences on channel choice, as well as technology choice, as indicated above. As an example, whereas China generally forbade foreign investment only a few years ago, joint ventures are now required to have a minimum 25 percent foreign partner share, and wholly owned subsidiaries are not only permitted, but to some extent encouraged as sources of capital and management expertise. Many firms also prefer wholly owned affiliates in order to enhance control, speed decision making, and to better protect proprietary technology. Consequently, the wholly owned subsidiary was the transfer mechanism of choice for firms in technologically dynamic industries. All electronics affiliates in the study were wholly owned subsidiaries. Food processors in China used joint venture forms with minority or co-ownership positions. One general manager of an experienced transnational food processor commented emphatically that the wholly owned form is always preferable. The perceived contribution of a local partner was relatively insignificant compared to the impediment to the transfer decision-making process implied in shared control.

In general, concern for government stability, proprietary property protection, and the felt need for control of the transfer process contributed to the propensity to transfer internally.

However, exclusive supply contracts were used uniformly by major firms to source branded athletic footwear and apparel. In China, manufacturers were joint ventures between municipal or provincial enterprises and Taiwanese firms having a long relationship supplying U.S. brands. Taiwanese managers provided a pivotal intermediary role, perceived as a cultural compatriots in the host environment, and as reliable partners by U.S. buyers.

Joint ventures were created to receive transferred technology, rather than being formed on the basis of the prior complementary skills existing in a potential foreign partner. In this study, the lack of prior experience with the transferred technology may indicate the dominance of start-up ventures in a variety of environments rather than a transfer to an existing foreign firm. In some situations, joint ventures were formed or reformed following extended periods of frustration with government bureaucracy or government-owned partners. Cross-cultural communication difficulty and differing perceptions of quality, equity, and responsibility seemed to play pivotal roles in these structural changes.

PROPENSITY TO TRANSFER TECHNOLOGY INTERNATIONALLY

The propensity of a firm's management team to transfer whatever technology via whatever mechanism should, first of all, be considered a dynamic, perhaps organic, process, contingent on the effectiveness of environmental scanning, responsive to new information, subject to bargaining, dependent on perceptions conditioned by culture. Consequently, by far the largest number of variables converge on this foundational element of the model.

In this region of the model, Robinson has aptly conditioned the assessment of cost and risk on perception. As a consequence, cross-cultural influence is seen not only in the installation of a new technology in a host country environment, but also in the supplier's decision-making process itself. Few concepts are more culturally conditioned than risk aversion and cost-benefit trade-offs over time.

The principal question concerns the extent to which the model describes the appropriate criteria, and their interrelationships, influencing the propensity to transfer technology internationally. Secondarily, how is that decision process moderated by the choices of technology and the transfer mechanism?

For overseas manufacturing cost centers, virtually all motivators to transfer translated into overall cost reduction. As one responsible liaison manager characterized the situation: altruism was a motivation only in the form of tangible economic incentives.

Although outside the scope of this chapter, in an aside, one manager commented that his firm was also involved in European production to transcend common market tariff barriers, despite higher labor costs.

In transferring manufacturing technology abroad, market penetration objectives were decisively long term rather than short term. Firms in the study had a clear sense that making overseas commitments and ultimate access to market development was in the context of their overall long-range global strategy.

The heavily bureaucratic and often confusing government decision-making process in China was perceived to be a constraint on technology transfer by subject firms. The Hong Kong and Singaporean government decision-making processes were not considered deterrents. In the latter case, though the government was perceived as authoritarian, the rules of the game were nonetheless clear and understandable, and the overall policy was viewed as supportive.

Foreign exchange availability was perceived as a constraint on transferring technology to China, but not at all to Hong Kong and Singapore. Firms in the study either export as a joint venture to gain foreign exchange or take payment in product.

Perceived government stability and integrity were highly correlated and virtually bimodal. Particularly in the wake of the Tiananmen Square incident, China was viewed as having substantially less stability and integrity than either Hong Kong or Singapore.

Infrastructure adequacy to support the transferred technology was perceived as a constraint in China. The availability and reliability of energy are a good example. When asked about a small building at the end of a manufacturing complex, host managers explained that it was an auxiliary power supply for the two operating days each week that the government did not supply electricity. Each firm in the area was allocated four day's power on a rotating basis. In Guangzhou, rather than cope with the uncertainty of energy availability, another firm took over management of a city power station. Improvements in the management and consequent efficiency of production and allocation benefitted not only the firm, but all other customers as well.

SUMMARY

Although no generalizations should be implied by cases in specific geographic and industry contexts, these emerging patterns do suggest support for the model in this composite situation. As the review of technology transfer model development revealed above, no known set of relationships approaches the Robinson model in terms of its comprehensiveness in the international context. The sheer magnitude of the number of variables, and the interactive nature of many of them, pose formidable barriers to measurement.

As anticipated, some elements in the general model were not found to be compelling for this particular mix of firms in their particular environments. However, the model was found to be useful in describing the structure among the profusion of variables influencing technology and transfer mechanism choice. Because the companies studied had already demonstrated a propensity to transfer technology internationally under specific conditions, the model template was also useful in confirming the original transfer decision. The analytical framework of the model was also effective in evaluating environmental and corporate changes that had taken place since the initial decision to transfer technology overseas, and to confirm the rationale to continue established relationships.

Notes to Chapter 2

1. D. B. Yoffie, Managing International Trade: New Dimensions in Competitive and National Strategies. Seminar presentation to the 25th reunion class, Harvard Business School, Boston, 1988.

2. P. J. Buckley, "A Critical View of Theories of the Multinational Enterprise." In P. J. Buckley, and M. Casson, eds.: *The Economic Theory of the Multinational Enterprise* (London: Macmillan, 1985).

3. Ibid, p. 2.

4. M. C. Casson, *The Entrepreneur: An Economic Theory* (Oxford: Martin Robertson, 1982).

5. K. E. Boulding, *Beasts, Ballads, and Bouldingisms* (New Brunswick, NJ: Transaction Books, 1980).

6. N. T. Wang, *China's Modernization and Transnational Corporations* (Lexington, MA: Lexington Books, 1984).

7. E. F. Schumacher, *Small Is Beautiful* (New York: Harper and Row, 1973).

8. J. Baranson, *Technology and the Multinationals* (Lexington, MA: Lexington Books, 1978).

9. H. Singer, *Technologies for Basic Needs* (Geneva: International Labour Organisation, 1982).

10. R. D. Robinson, *The International Transfer of Technology: Theory, Issues, and Practice* (Cambridge, MA: Ballinger, 1988).

11. B. L. Kedia, and R. S. Bhagat, "Cultural Constraints on Transfer of Technology Across Nations: Implications for Research in International and Comparative Management," *Academy of Management Review*, 13(4) (1988), 559–71.

12. M. Z. Brooke, *Selling Management Services Contracts in International Business* (London: Holt, Rinehart and Winston, cited in Robinson, 1988), p. 4.

13. Robinson, pp. 4–5.

14. Ibid.

15. Ibid.

16. J. Shumpeter, *The Theory of Economic Development* (Cambridge, MA: Harvard University Press, 1934).

17. D. G. Marquis, and T. J. Allen, "Communication Patterns in Technology Transfer," *American Psychologist*, 60, (1966), 1053.

18. W. H. Gruber, and D. G. Marquis, "Research on the Human Factor in the Transfer of Technology." In W. H. Gruber and D. G. Marquis, eds.: *Factors in the Transfer of Technology* (Cambridge, MA: MIT Press, 1969), p. 256.

19. C. H. Smith, *Japanese Technology Transfer to Brazil* (Ann Arbor, MI: UMI Research Press, 1981), p. 16.

20. Gruber and Marquis, p. 257.

21. Ibid., p. 262.

22. Robinson, pp. 54, 79.

23. E. M. Rogers, "Key Concepts and Models." In R. A. Solo and E. M. Rogers, eds.: *Inducing Technological Change for Economic Growth and Development* (East Lansing: Michigan State University Press, 1972), p. 95.

24. R. A. Solo, "Technology Transfer: A Universal Process." In Solo and Rogers, p. 7.

25. R. E. Brasseur, "Constraints in the Transfer of Knowledge," *Focus,* 3 (1976), 16–18.

26. S. N. Bar-Zakey, "Technology Transfer Model," *Technological Forecasting and Social Change,* 2(3/4) (1971), 323.

27. J. Baranson, *Technology and the Multinationals: Corporate Strategies in a Changing World Economy* (Lexington, MA: Lexington Books, 1978).

28. R. A. Burgleman, and M. A. Maidique, *Strategic Management of Technology and Innovation* (Homewood, IL: Irwin, 1988).

29. M. D. Harvey, "Application of Technology Life Cycles to Technology Transfers," *Journal of Business Strategy,* 5(2) (1984), 51–58.

30. T. S. Robertson, and H. Gatignon, "The Diffusion of High Technology Innovations." In J. M. Pennings, and A. Buitendam, eds.: *New Technology as Organizational Innovation* (Cambridge, MA: Ballinger, 1987), 179–96.

31. J. M. Pennings, "Technological Innovations in Manufacturing." In Pennings and Buitendam, pp. 197–216.

32. J. W. Creighton, J. A. Jolly, and T. A. Buckles, "The Manager's Role in Technology Transfer," *Journal of Technology Transfer,* 10(1) (1985), 65–80.

33. A. C. Samli, *Technology Transfer* (Westport, CT: Quorum Books, 1985).

34. Kedia and Bhagat.

35. W. F. Yager, Transnational Corporate Strategy and Economic Development: A Synthesis, unpublished paper, University of Oregon Graduate School of Management, 1986.

36. Ibid.

Chapter Three

Interfirm Technology Transfers and the Theory of Multinational Enterprise

Farok J. Contractor

INTRODUCTION

Whereas the bulk of commercial international technology transfers occur intrafirm, that is, within the multinational firm, the share of interfirm transfers may have grown in recent years. Or at the least, we may say that technology transfers between independent and quasi-independent international companies remain a significant economic phenomenon.

This chapter examines the choice between intrafirm and interfirm transfers. It begins at a point where a multinational firm has already developed a technology and now seeks to exploit it in many, if not all, national markets. In some regulated economies the choice may not exist, of course. The host government may mandate the transfer and sharing of expertise with local firms as a condition of doing business. For this chapter we are concerned with the more interesting case that exists in the majority of industrialized nations where fully owned operations are permitted to foreign investors (in most industries) and where the full spectrum of ownership choices are available to investors. Given the possibility of full ownership, why do some firms opt for the alternative of transferring and sharing technology with corporate partners in foreign countries?

The question is a crucial one for the theory of the multinational enterprise because an important leg of the theory from Hymer to Dunning[1] remains the ownership or firm-specific advantages that the firm carries across national borders. Under what circumstances is the sharing of technology with other firms a substitute for the multinational enterprise extending its own organization across national borders?

Second, this study examines changes in the global economy and company strategies that increase the marginal propensity to share technology with other international companies. Whether such a marginal shift has occurred remains an open empirical issue. Certainly, the business press in the 1980s reported the formation of a huge number of international joint ventures, technology-sharing, and licensing arrangements. The important point is that the bulk of these have been formed in the Organization for Economic Cooperation and Development (OECD) nations where the option of establishing fully owned operations is usually available.

In this chapter the term "technology" is used broadly. It is not confined to engineering or production, although this is most common, but it may include proprietary marketing or administrative techniques sufficiently differentiated so as to constitute a firm-specific advantage. For instance, there is an active international "market" in Strategic Information Systems involving transfer of U.S. technology under licenses and joint ventures. The word "market" is used advisedly, for the very idea of differentiated, firm-specific knowledge is partially incompatible with the traditional notion of a market. Indeed, some early economists went so far as to treat technology as akin to a public good and the very attempt to market it as suboptimal, if not a gross distortion. Granting that research and development expenditures needed to be recompensed in order to provide an incentive for private investment, Johnson wrote: ". . . once new knowledge has been created, it has the character of a public good . . . so that optimality requires that it be made available to all potential users without charge."[2]

Of course, although pure knowledge has the characteristics of a public good, knowledge is only one of five distinct elements of a technology transfer agreement that can include some or all of the following:

1. Services such as training and installation.
2. Information that results in knowledge.
3. Legal rights to use this information or symbols (trademarks) in certain territories.
4. Legal or de facto restraints placed on the technology buyer.
5. Objects such as instrumentation and equipment containing encoded information.

Such a diverse collection cannot be marketed easily. There are valuation difficulties, bounded rationality on part of the negotiators, information asymmetery, and in general high transactions cost that make the market for international interfirm transfers imperfect. The object of this chapter is not so much to detail these imperfections as to point out the circumstances under which they may be overcome. And overcome they must be, empirically speaking. According to Contractor and Lorange,[3] even before the 1980s surge in joint ventures, American companies had extant over 30,000 current arms-length licensing agreements and about 13,000 quasi-arms-length relationships in the form of minority and 50–50 foreign affiliates; this was as many as the majority plus fully owned affiliates put together. In addition, according to Oman[4] and Franko,[5] "new forms" of technology transfer such as management service contracts, turnkey agreements, and coproduction proliferated in the late 1970s and 1980s.

Outline of the Study

In seeking the causes of interfirm technology transfers, this chapter treats both the theory of the multinational enterprise (much of which seeks to explain why the firm will *not* transfer technology to other firms) as well as conditions in the global environment that provide exceptions to this theoretical prescription.

1. As a brief introduction, it is worth reviewing for readers indicators that illustrate the erosion in the relative position of the United States in world technology only to show how interfirm transfers are more likely in a multipolar world.
2. Next we ask what the theory of the multinational enterprise has to say about interfirm transfers as an alternative to internalization of firm-specific advantages.
3. Third, we examine attributes of a technology that may make it more transferable, that is, are some technologies more transferable than others?
4. Finally, the focus is on other changes in the global business environment over the last two decades that foster the greater use of interfirm transfers over the transfer of technology within the multinational firm.

THE UNITED STATES IN A MULTIPOLAR WORLD

The salient international economic change of the last quarter-century is the erosion in the relative (not absolute) position of the U.S. economy vis-a-vis other nations. It is one thing to prefer fully owned subsidiaries and disdain joint ventures and licensing when one's technology confers a strong bargaining position and market lead, when competition is relatively weak, alternative global sources of technology are few, and when the transactions costs of technology transfer to relatively backward recipients outweigh potential returns from external agreements. Such assumptions may no longer hold true in many industries. Even the archexponent of fully owned foreign subsidiaries, IBM, changed its corporate policies in the 1980s to allow joint ventures and licensing in selected cases. By comparison, multinationals based outside the United States have historically had a higher propensity to engage in interfirm agreements.

The new U.S. position is commonly illustrated by citing the change in the American share of world industrial output from 49 percent after World War II to 17 percent today. More pertinent are indicators such as the change in the share of U.S. patents, now down to a quarter of the world total. For the purpose of this study, patents filed outside the country of the applicant are more relevant. By this criterion, U.S. applications abroad, which in 1969 equaled those of West Germany, the United Kingdom, Japan, and France combined, now constitute about half of those four countries' external filings. In the United States, "foreign residents" are now issued almost half the patents annually; Japanese residents alone account for an 18 percent share of the U.S. total.

Another index is a country's ratio of exports over imports for the so-called high technology manufactured products. In 1982, the U.S. ratio was 1.25 compared with West Germany's 1.20, France and the U.K.'s 1.05, and Japan's 3.00. More ominously, since 1985, this ratio for the United States has sunk below 1.00.[6]

The purpose here is not to debate the validity of these indicators or their causes. There exists a large literature on the meaning of patent statistics and research expenditures, as exemplified by Glismann and Horn,[7] or Mansfield.[8] Here we simply wish to illustrate how companies are in a far more multipolar strategy landscape then 20 years ago.

In constant dollar terms, despite the increases in R&D initiated under the Carter administration, the U.S. share of R&D expenditures fell from 57 percent to 53 percent of the five nation total for France, the United Kingdom, West Germany, Japan, and the United States.[9] But this modest decline belies the fact that after subtracting federal R&D, mostly defense-related, nonfederal R&D expenditures rose by only about $4 billion between 1975 and 1981 (in constant 1972 dollars). In relation to the size of the economy, U.S. R&D expenditure as a percentage of the GNP is now about the same as Japan, the U.K., and West Germany.

However, because these other nations spend proportionately less on defense R&D, their nondefense companies spend a somewhat higher percentage than U.S. companies on average. It is because the United States is so huge an economy that its total figures on R&D expenditures outdistance any other individual OECD nation.

The size of the U.S. market continues to confer an advantage to companies in terms of both economies of production, as well as a broader base of sales revenues from which to amortize R&D outlays. However, in some industries such as semiconductors, computers, and telecommunications, succeeding generations of technology have gotten so expensive (as product cycles have shrunk) that no one market or even a region is large enough now. In some applications, R&D is now so expensive that is can be justified only by the hope of a return from a global market. The risk may be so high that even industry giants such as AT&T or Fujitsu will form joint ventures, partially motivated by the market expansion and R&D risk-sharing contribution their partners will make.[10]

INTERFIRM TECHNOLOGY TRANSFERS AND THE THEORY OF THE MULTINATIONAL ENTERPRISE

Theories explaining why a firm would internalize its proprietary advantages by retaining technology within its hierarchical control make three principal arguments. The first has to do with the "appropriability" of profits or rents in foreign markets. The second focuses on imperfections in the market for international technology and posits that if imperfections are great, then the firm will prefer to exploit technology via its own affiliates rather than share or sell it to other companies. The third, which is a subset of the market imperfection argument, treats transactions costs. Let us consider each in turn.

Appropriability Theory

The "appropriability" argument pioneered by Hymer[11] (1960), and amplified by Caves[12] and Magee[13] focuses on the monopolistic revenue-earning power of multinational enterprise (MNE)-made products. Keeping their technology, marketing skills, patents, and trademarks internal, MNE products are said to command higher prices than their local competitors, if any, in the host market. A second variation of the appropriability theory is that not sharing administrative decisions and control with local entities leaves the MNE free to set the monopolisitically optimum price and "appropriate" the maximum return on assets. Thus one strand of the argument simply states the MNE products will command higher rents. The second strand deals with the appropriation of profits unhindered by agreement-bound formulae (such as a licensing fee or royalty) or by the divergent proclivities of joint venture partners.

For technology transfer to take place, the multinational firm must calculate as high a *net* (risk-adjusted) return from a license or a joint venture as it would receive from a fully owned affiliate. This study is concerned with outlining the conditions under which this is possible. With competent local firms it is increasingly feasible for them to manufacture products equal in quality to the MNE's standards—and command as high a price as the MNE itself would have charged, especially if the international trademark is also transferred. The second postulate is more difficult (though by no means impossible) to overcome. A licensee or joint venture partner means that the rent is now to be shared, rather than captured by only one firm. However, this assumes a static rent, invariant to the ownership variable. As much of the recent literature implies, the very presence of a partner may expand the market to such an extent that the MNE is better off even after sharing its proceeds.[14] This is the basis for strategic alliances between multinationals, which could have chosen to compete instead. Moreover, the bilateral division of rent between the technology supplier and recipient could be such that the former captures the return on technology, whereas the recipient merely earns a return on capital invested.

Let us introduce a deliberately provocative example. Suppose a small American research laboratory has patented an important drug and licenses it in Mexico. The Mexican license may be content to earn a little more than the normal return on capital investments, whereas the American firm extracts a generous return on its technology by earning a royalty of, say, 40 percent on sales. This is feasible because gross margins on some drugs can easily exceed 50 percent. This arrangement could well be a positive-sum game. Being a small biotechnology company, the technology supplier lacks market reach and capital. To them, such a royalty arrangement can earn as much or more than the net return from a direct investment. The Mexican recipient has a medical distribution network and can add a product line at little incremental cost. We know that some mutually beneficial arrangements of this kind are increasingly undertaken. But many are not, because of "market imperfections," discussed next.

Imperfections in the International Technology Market

A second branch of MNE theory deals with failure and inefficiencies in the global market for intermediate inputs. When considering technology transfer negotiations, proprietary company information, techniques, and procedures sometimes cannot be easily codified into manuals, blueprints, or into a tangible form that a prospective buyer can assess and price. Much will reside in the thinking of key administrative and technical personnel whose relationships determine the rate of diffusion of technology within and across organizations. Technology transfer is usually not to be viewed as an act but as a relationship overtime whereby the technology recipient learns from organizational (that is, human) links with the supplier company.[15] Of course, in some industries, technical information is sufficiently codified in the patent and/or prospective licensees sufficiently capable technologically, that the mere act of conveying the legal right or permission to the licensee is sufficient. But this is rare. Usually, a sufficient degree of unpatented and undisclosed but proprietary information must accompany the patent, if any, to render the package worthwhile to the receiver. But the technology "sellers" cannot reveal too much beyond the publicly available patent (usually a bare-bones statement) without giving away vital information to prospective "buyers." Buyers therefore have to operate in uncertainty and on faith. This makes the technology market inefficient. This means that a MNE offering a technology package may not be able to command as high a price or return, as compared with the MNE exploiting the technology on its own in foreign markets.

Other impediments to the technology market stem from host government interventions or simply tradition. In the above example, even through the American company, Mexican licensee, and consumers may benefit, a royalty rate of 40 percent may be considered shockingly unurious. In technology licensing, conventional rates are well below 10 percent and average under 5 percent. The Mexican government moreover puts ceilings on such rates.

However, whereas substantial imperfections and impediments remain, these are also overcome as technology becomes increasingly codified (that is, increasingly transferable and assessable to prospective buyers), as technology brokers and consultants proliferate, and as recipients in all parts of the world are better able to absorb technology with fewer of the training and attendant services compared to decades past. We explore these themes in further detail later. Suffice it to say here that these changes have marginally lowered uncertainty and transaction costs in the international technology market.

Transaction Costs in Interfirm Technology Transfers

The net return from a technology transfer transaction is the gross return less the transaction costs. If transaction costs are high, then an introfirm transfer is preferred. Coase[16] and Williamson[17] have their ideas applied to the multinational firm by Hennart,[18] Anderson and Gatignon,[19] and others.

As we see, transaction costs in international technology agreements need not be a formidable barrier. Moreover, the literature neglects the fact that intrafirm transfers across national boundaries also entail costs; furthermore, setting up a controlled foreign affiliate entails additional ongoing governance costs, which may not be adequately measured because these are part of MNE headquarters overhead. The point we arrive at is that transaction costs may in fact be low and not pose a large barrier in many cases.

Be that as it may, let us examine the three components of transaction costs as they relate to technology transfers between international firms. First, there are costs of negotiating and transferring the information and capability to the other firm and training their personnel. These can be sizeable, especially if the receiving firm is unsophisticated. The total negotiation and transfer costs can be estimated quite accurately. Second, there are the opportunity costs of abdicating a market in favor of the licensee or joint venture corporation. This is measured by estimating the foregone direct sales and profits that the firm could have made on its own in the market, had it not relinquished its presence there under the agreement. This is, strictly speaking, not a transaction cost, but a cost that results from the transaction. We may note that whereas there are nonexclusive grants of a technology, in practice technology transfer often entails the partial or complete vacation of a market or a product line in favor of the technology receiving company. The third "transaction" cost is the threat of creating a competitor in markets *beyond* the purview of the agreement, or beyond the anticipated life of the arrangement. This is difficult to estimate because it entails a judgment of the future likelihood of partners going their separate ways, the degree of assimilation of technology by them, and their competitive position in other markets. This uncertainty and worry can turn into a virtual phobia about licensing and cooperative ventures with Japanese firms, exhibited by many U.S. firms, and in writings such as Abegglen.[20] Japanese firms with their technological sophistication can indeed become formidable competitors. But U.S.-Japanese joint ventures are perhaps at an all-time high.[21] It is too much to believe that so many American executives are naive, or short-term profit-oriented, as the popular press and even scholars allege. They must have carefully assessed the risks of potential competition and contained these by either agreement-stipulated or strategy-imbedded provisions, discussed later. Certainly, they must have considered the benefits to outweigh the potential risks. Moreover, in developing and socialist nations the actual propensity and likelihood of competition from licensees and partners outside their home nation is low.

To summarize, three types of transactions costs can exist in technology transfers. These are the negotiations and transfer costs, the opportunity cost of foregone direct business, and finally the risk of creating a future competitor. However, transfer costs need not always be high, and we should not forget that there are costs in transferring technology to an affiliate as well (although these would probably be lower). The second, opportunity cost may also exist in establishing a fully owned subsidiary. For instance, direct export sales to a nation and the profit accruing thereon may be suspended once an affiliate is set up for local production. Finally, the risk of creating a competitor is often exaggerated except in technology transfers to Japanese companies. Even there, with effective patents and other provisions, the recipient can be confined to designated territories.

ATTRIBUTES OF TECHNOLOGY THAT FOSTER
INTERFIRM TRANSFERS

In later sections we discuss changes in the international business environment that have marginally shifted MNE choices in favor of technology transactions over fully owned direct investments. First let us examine three related attributes of technology through which the environment impinges upon the business decision—transferability, separability, and the extent of codification.

In recent decades, the transferability of technology has increased, as intended recipients of technology in the industrialized nations and the NICs have become more sophisticated, requiring less training and start-up assistance than before. In a cross-sectional analysis of 33 advanced and developing nations, that the proportion of supporting technical services in the overall technology package dropped, the higher the per capita GNP of the recipient's nation.[22] With appropriate caveats, this finding supports the idea of higher technical absorptive capacity (that is, lower transactions costs) as a nation, and the world industrializies.

Separability refers to the ability to unbundle the elements of the technology and assemble a package specifically needed in the recipient enterprise. To the extent that this can be done, it increases the likelihood of a transation in those cases where the entire package may be too expensive for the purchaser. Separability is a function of the maturity of the technology and the ability of the recipient firm to manage without some elements of the package. Most importantly, technology suppliers in some cases are more willing to unbundle than in the past. The reason is not altruism, nor is it caving in to this long-standing demand on the part of developing nations. There is money in it. Two examples follow. In the past, American can companies preferred to put up their own can producing factories abroad, or at most, supply such a factory on a turnkey basis. Each such factory might involve several score technologies from metal drawing to anodizing. Nowadays, some are willing to spin off particular technologies such as just the anodizing process for a fee. Not only are there anodizing applications in a number of other industries, thus increasing the pool of possible technology buyers, but there is little change that the disclosure of the anodizing applications *alone* will significantly erode the overall technical lead and expertise of the can company and create a competitive threat. Similarly, a steel company may freely license technology for certain ovens, even if it is unwilling to supply technology for its overall operation.

In any large company there are likely to be hundreds of separable technologies, many having applications in unrelated industries. Several companies in the 1980s have discovered their transfer value and have set up "technology marketing groups" to inventory and offer technical assets for sale.[23]

The willingness to separate pieces of the production and value-added chain and have these functions performed externally is part of what appears to be a de-integration trend in several U.S. industries. For example, Toyota and Nissan have 18 employees per thousand vehicles made, compared to over 100 employees at General Motors in the early 1980s[24]—a difference principally explained by the number of external suppliers. The value-added over sales ratio was 0.48 for General Motors but only .15–.20 in Japan. In recent years GM has been switching to a policy of taking an equity stake in some suppliers and transferring technology to them, rather than relying on a policy of high vertical integration. They now have over 200 joint venture relationships. In Japan, of course, there are intense transfers of technology both ways from auto company to its "Kai" (Association of Cooperating Suppliers), and vice versa. Such relationships are also becoming international as automobile manufacturing becomes more global.

This chapter does not pursue the question of the optimum level of integration in the automobile industry. All we are pointing out is that a marginal de-integration is taking place in this and some other industries. The de-integrating firm has to separate its technology into discrete pieces and transfer them to partners.

The third attribute of technology that affects its transferability is codification. All technology begins as tacit knowledge but is increasingly codified as the technology cycle progress.[25]

Codification refers to the ability to reduce the technical information to manuals, programs, and written procedures, as opposed to transferring it via human expertise and training. In part, this depends on the technology, of course. But it also depends on the ability of the receiving firm's personnel, which has improved over time, and the spread of computerized expert systems and "intelligent" machines. Technology transfer cost is thus reduced. Moreover the pool of potential technology buyers is thereby increased.

CHANGES IN THE INTERNATIONAL BUSINESS ENVIRONMENT

As the transferability of technology increases and as transfer costs decline, the technology market becomes more efficient. One has only to examine the changes over the past quarter-century in the numbers of scientists and engineers in developing nations, the middle-income countries, as well as traditional large technology buyers like Japan to realize by how much the pool of buyers has grown. Moreover, in many nations, government policies are skewed in favor of market transaction-based import of technology. For three decades after World War II, Japanese industrial policy excluded or discouraged inward foreign investment in the shape of fully owned subsidiaries.[26] It encouraged licensing and joint ventures, judging these to be cheaper means of technology acquisition for the nation as well as quicker internal diffusion technology than fully owned foreign investments. This past Japanese policy is a model that several developing nations are trying to emulate. However, a caveat is in order. Several nations' attempts to implement such policies have been without much success because their internal technological capabilities have not yet grown strong enough to be able to assimilate sophisticated technologies without the full commitment and long-term presence of MNEs. To get that commitment often entails allowing them to set up fully owned and controlled affiliates. We return to this theme when discussing government policy.

Risk

The proposition advanced here is that in industries and countries characterized by higher investment risk, there is a marginal shift toward the agreement-based modes. We advance the hypothesis that the 1980s have constituted an era of higher investment risk in general compared to past decades. This hypothesis has to be viewed with considerable caution. If by risk we simply mean variance in profits, then there is some evidence to show that the 1980s have indeed been more volatile with respect to foreign exchange rates, inflation, and interest rates than in decades past, contributing to fluctuations in multinational company profits.

But would agreement-based modes of international business not suffer the same volatility? No. In general these are more immune to business cycle and inflation variance because most of the compensation formulae in licensing, management-service contracts, assembly, contract production, and, sometimes, even in joint development, link the earnings of the technology "seller" to a sales figure, rather than a percentage of profits. And it is axiomatic that sales fluctuate far less than profits over the business cycle. No research has been done on this issue, apart from one or two case studies such as Contractor, who relates instances of American companies whose foreign profits dried up during the global recession of the early 1980s, whereas their royalty income continued to flow at almost normal levels.[27]

Of course, the above argument does not apply to equity joint ventures where the principals share the profits of the venture. But even these are often accompanied by royalty arrangements, lump-sum payments, component supply, and buyback, which are not linked to profits but rather sales.

The 1980s have also been characterized by an increase in real interest rates. This affects the corporate strategy choice at the margin, as illustrated in Figure 3.1. Let us suppose that a company has a choice of making a fully owned investment (I) or an agreement-based technology transfer (L) where the returns are linked to the sales level of the recipient enterprise. The figure plots the reltionship between net present value (NPV) and the discount rate (r). For normal projects, $\partial(NPV)/\partial r$ is negative. At low discount rates, the net present value of the fully owned investment is often higher than the external technology transfer alternative. $NPV_I > NPV_L$. The internal transfer choice will be preferred.

However, because of the differrent chronological asymmetries between the cost and income flows in the two alternatives, usually we find that

$$\left[\frac{\partial(NPV)}{\partial r} \right]_I > \left[\frac{\partial(NPV)}{\partial r} \right]_L$$

The slope of the investment function is steeper than that of the agreement alternative. This means that as a company applies a higher interest or discount rate to project cash flows of the two alternatives, their NPV will break even and cross over.

Figure 3.1 shows the breakeven discount rate as (A). However, that assumes the company is using the same companywide discount rate (A) to evaluate both alternatives. This is not appropriate. Suppose a company used a base discount rate (C). Because agreement-bound transfers, especially those where the compensation is linked to sales, are *inherently* less volatile and less subject to political and currency-conversion risk, a lower rate $(C - r_L)$ may be used. In contrast, for the investment alternative, companies will often apply a risk premium r_I for foreign investment and use a higher discount rate $(C + r_I)$ than the company average. In doing so, the breakeven discount rate moves from (A) to (C).

The basic proposition is simply that there is often a breakeven or crossover point; that in an ear of higher real interest rates, *some* companies' strategy choice will switch; and that in the aggregate sense, there will be a *marginal* shift from investment to other technology transfer modes such as licensing as a result of this change in the economic environment.

We might enumerate, in passing, other corporate risk factors that are inducing a greater degree of interfirm cooperation. We have mentioned one: R&D costs accelerating with each generation of technology in some industries, as product cycles shrink. The risk may become so high that even large MNEs: (1) join hands to share the developmental risk, (2) have a larger joint market over which to amortize the costs, and (3) by reducing competition perhaps even prolong the product cycle for each generation of technology. Other benefits accrue form a lower capital investment, for example, fewer

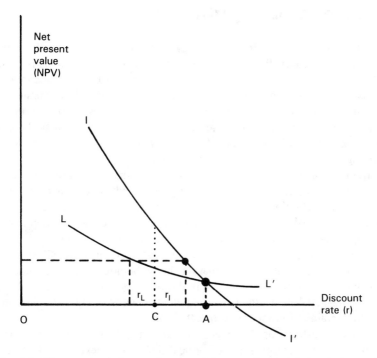

Figure 3.1. The strategy choice as a function of the cost of capital facing the international firm.

assets are at risk in joint ventures compared with an equivalent-capacity, fully owned installation. This is because of joint economies of scale, rationalization, and pooled market demand, and utilization of slack fixed assets from the parent companies.[28] In the pharmaceutical or biotechnology area, risk is lowered by faster entry and government certification, as a result of swapping the technology and clinical testing data of the principals in the venture. Licensing and joint ventures can often be a lower cost means of testing a foreign market considered too risky for a full investment. Moreover such agreements, especially those involving local interests, are viewed more favorably by the host government and bear a lower political risk. In general, companies in high technology industries diversify their technology portfolio and reduce commercial or political risk by making partial investments in several ventures rather than using the same capital over fewer fully owned investments. In industries characterized by rapid change, uncertainty, and many alternative technologies, this means that a company increases the probability of having a claim to a promising new development generated by one of the ventures in which it has a stake; see Harrigan and Contractor and Lorange for futher discussion.[29]

An Emerging Technology Market

An international market for technology transfers organized in the 1980s. There are technology lists, advertisements, notices, and data banks under the aegis of professional engineering and licensing executive societies, as well as multilateral agencies such as UNIDO. The acid test of whether such technology markets actually work is whether the business will support middlemen and brokers. Indeed, there have emerged a number of brokers and consulting organizations, some of whom organize their own expositions, attracting buyers and sellers from several nations.

Perhaps even more telling is the fact that major corporation have established internally "technology marketing groups" whose task is to: (1) identify the many technologies strewn throughout a large company, (2) negotiate with product divisions and/or the international division which technologies may safely be offered to outsiders and which are to be reserved for the company's own use, (3) identify potential customers in other industries and nations and in general, and (4) market the technologies. In this endeavor, the technology-selling firm seeks to multiply the potential number of buyers while taking legal measures (such as patent registrations) or strategic measures to contain the competitive threat, if any.

In some industries rapid changes in market applications induce interfirm cooperation. Take the example of optical disks. The developer is typically a large computer-technology company with market experience limited to business or home data processing needs. But optical disks have *applications* in a great many fields such as publishing, music, video, and so on. Hence the justification for joint ventures between the computer company, which cannot afford to keep pace with all the downstream market applications, and companies with market knowledge in each application. This point takes on added force when we add the international dimension with separate country markets.

Changes in Corporate Culture and Policy

The traditional (and in many companies, continuing) preference of executives is to expand internationally by fully owned affiliates that reproduce corporate cultures in foreign settings. But this option is simply unavailable in many resulted developing countries, as well as in OECD nations such as France, Japan, or Canada, which protect key sectors such as telecommunication or bib-sciences, by requiring joint ventures with local firms.

In selected circumstances, however, companies are dicovering that it makes sense to link up with a local partner even when there is no external or governmental mandate to do so. This is the key proposition of this chapter. Joint ventures, strategic alliances, licensing, and other nontraditional modes of business are being formally factored into corporate planning and strategy. They are no longer always considered to be "second best" to the internalization strategy. The propensity to do so appears to be relatively higher among diversified MNEs, medium to small-size firms, and culturally adaptable organizations.

Take a diversified giant like General Electric. With hundreds of products and sales in scores of nations, the number of product/market combinations approaches 10,000. There is no way even a large company can meet all of these combintions from fully owned plants because of financial or personnel constraints. In their planning review, what review companies are doing is to assess the technology transfer question for each combination of

- Technology
- Product
- Market
- Piece of the value-added chain

For the most vital combinations, the "crown jewels," internalization is generally preferred, if environmental and regulatory conditions allow. For the rest, there is greater flexibility and willingness to accept partners and licensees than in years past, in order to expand the applications of the company's technologies to other nations and industries rather than foregoing business as in the past.

The new flexibility is the result of greater corporate sophistication on three related issues: (1) the ability to manage multicultural enterprises, (2) the ability to delink control from ownership, and (3) containing the competitive potential of partners and licensees. The growing sophistication on cultural matters, especially among U.S.-based firms, which have hitherto been insular, is covered in the growing literature on cross-cultural management[30] and its measurement.[31] In organizational development terms, Child and Keiser[32] examine internal adaptation to facilitate the absorption and transmission of technology. Anderson and Gatignon advance the hypothesis that the greater the cultural distance between the two international firms, the more efficient agreement-based entry modes involving "low control" may be because of low transactions cost.[33]

Nonetheless, it is possible to wield control by means other than the formal content of an agreement.

Control can be exercised by sharing only part of the process as in the case of Boeing's corporation agreements with Japanese firms, which receive the technology for only parts of the aircraft. Control can be exercised by keeping a license or joint venture dependent on one of the parents for a critical component, an active ingredient (for example, in pharmaceutical licensing), or for access to important markets that one of the partners controls.

The competitive proclivities of licensees and partners can be restrained with perfect legal sanction when the technology supplier maintains strong patents and trademarks registered in all important territories. Moreover, this enables agreements to be signed separately in each viable market. Even if *each* agreement yields only a small profit over its typical five-to seven-year life, in the aggregate, over *all* agreements worldwide, the total can be sizeable enough to make a selected external technology policy very worthwhile for the MNE.

The future competitive potential of partners can thus be contained by maintaining mutual dependencies and mutual vulnerabilities. In the ultimate analysis, keeping one step ahead of competition by reinvesting a sufficient protion of earnings back into research is the best defense. Here, a judiciously selective use of external agreements not only augments current resources available for R&D, but in industries characterized by multiple spin-offs, technological fragmentation, or R&D costs too high for even large firms to bear alone, provides access to a range of technologies no one company can generate.

SUMMARY, CONCLUSIONS, AND GOVERNMENT POLICY IMPLICATIONS

The theory of the multinational enterprise posits that, due to higher revenues commanded by MNE products, and transaction costs and market imperfections in the interfirm technology market, it is usually more efficient for firms to use their own internal organization to serve foreign markets. However, relaxing and modifying these assumptions allow us to explain the existence of other agreement-based modes of international business where technology is shared with other firms. These modes include licensing, joint ventures of all types, consortia, coproduction agreements, and joint R&D, which appear to have increased in number during the past decade. This chapter details the changes in the environment over the past 35 years that foster cooperation and transfers of technology between international companies.

We have so far examined the choice between intrafirm and external technology transfers from the point of view of the company that has developed and owns a technology. Now we ask what is the host nation's viewpoint on the question of how technology is acquired by the country.

In some cases government policies have rendered the wholly owned subsidiary a more costly form of multinational investment while making a joint venture or licensing-based technology transfer a superior or optimal form of foreign investment. Examples range from LCDs giving more lucrative incentives to joint ventures with local interests, for example, in India; to the French government adopting technological standards in computers and telecommunications with the specific intent of keeping IBM and AT&T out, or at least to impose sufficient costs on them, thereby giving local firms an advantage.

Other examples of government policies helping the agreement-based modes of operation include technology data banks, whereby prospective purchasers of technology are better informed of global sources of supply, suggestions from ministries to get alterntive competing quotes, and negotiations assistance. But such policies always appropriate?

It is necessary to consider the drawbacks of government regulations, which are often contradictory. Although some regulations are designed to facilitate the agreement-based imports of technology to the nation, in operational terms they are in fact directed at improving the negotiation strength of local licensees or joint venture partners and to improve the terms of their "purchase." Hence in some nations they can have the opposite effect of retarding the flow of technology, if the country as a whole is thereby seen as unattractive and too restrictionist by foreign investors. There is an optimum level of intervention that most governments have not yet come to grips with.

A fundamental policy question is presented as hypotheses for future research in Figure 3.2. Are agreement-based modes always the best from the point of view of the nation? Fully owned foreign investments may carry the highest national costs, in terms of repatriated earnings, economic rent accruing to foreign investors, continued dependence on foreign technology and imports, and balance of payments outflows. But fully owned foreign investments may carry the highest benefits to the nation in the form of the latest technology, and administrative and marketing methods. In contrast, requiring local participation generally reduces some of the national costs and helps speed the diffusion of technology across firms within the nation (as opposed to the MNE internalizing its expertise). But the benefits may also be lower. Studying India's policy, which required local equity participation, Balasubramanyam found no evidence that it was good for India overall.[34] Much depends on the industry, the complexity of the product, existing levels of skills in the nation, its technological absorptive capacity, and so on.

The relative national benefits and costs of the mode of technology acquisition, ranging from purely arms-length purchases in licensing to intermediate forms of partnership such as joint ventures, to fully owned investments, vary by both *country* as well as *industry*. Figure 3.2 illustrates a hypothesis that only in the more industrialized nations, with a strong indigenous technical capability, might joint ventures and licensing confer higher

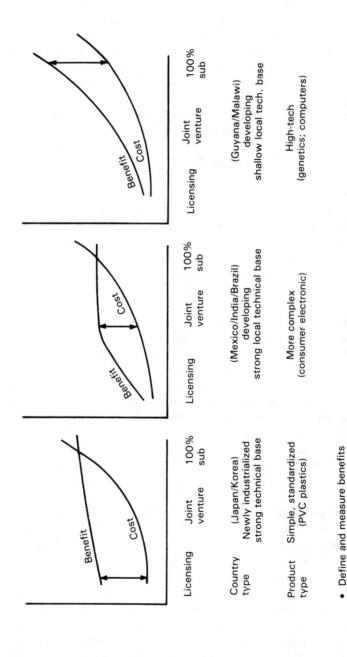

| | Licensing | Joint venture | 100% sub | Licensing | Joint venture | 100% sub | Licensing | Joint venture | 100% sub |

Country type   (Japan/Korea) Newly industrialized strong technical base   (Mexico/India/Brazil) developing strong local technical base   (Guyana/Malawi) developing shallow local tech. base

Product type   Simple, standardized (PVC plastics)   More complex (consumer electronic)   High-tech (genetics; computers)

- Define and measure benefits
- Define and measure costs   for each mode of foreign participation

Figure 3.2. Hypothesis for national entry policies: Should all nations follow similar policies? Same policy for all products?

net national benefits than fully owned foreign investments. Even there, for high technology items fully owned foreign investments are hypothesized to be superior.

Regrettably there are few studies on this issue, except for country-specific works such as Balasubramanyam (1973) for India[35] and Rhee and coworkers for Korea.[36] Hence the assertions in Figure 3.2 must remain as hypotheses. Suffice it to say that across-the-board prescriptions of investment policies, for several countries or within a country for all industries, are going to be suboptimal for the nation. This is in fact the conclusion reached from another perspective, by Encarnation and Wells.[37]

Finally, returning to the industrialized OECD nations, the 1980s saw a general relaxation in antitrust enforcement, especially in the United States where joint research consortia comprised of direct competitors are now actually encouraged to maintain national competitiveness.

Overall, it appears that the political and economic environment for international business operations manifests two divergent trends. In some industries there is a trend toward "globalization." This means that with converging buyer preferences and technical standards, MNEs produce a range of fairly standardized products for all markets from a few large "rationalized" and integrated global factories. In such a case there is a strong emphasis on efficiency and internal control and maintaining technical and administrative expertise within the firm. Cost reduction, technology, scale, and earnings extraction are concerns that dominate reponsiveness towards the local environment.

In other industries however, strong local customer preferences, different sourcing requirements, varying technical standards in different countries, economic nationalism, transport and tariff barriers, high R&D risks, and other factors detailed here point to sharing technology with other firms as a viable and sometimes superior option.

The sophisticated MNE will know when to share technology, and when not to. This chapter lays the theoretical basis for such discriminatiton.

Notes to Chapter 3

1. S. Hymer, *The International Operations of National Firms: A Study of Direct Foreign Investment*, Ph.D dissertation, MIT, 1960. (Cambridge, MA: MIT Press, 1976); and J. Dunning, "The Theory of International Production," *The International Trade Journal*, III, no. 1, (Fall 1988), pp. 21–66.

2. H. Johnson, "The Efficiency and Welfare Impliciations of the International Corporation." In C. P. ed.: *The Multinational Corporation*, (Cambridge, MA: MIT Press, 1970).

3. F. Contractor and P. Lorange, "Competition vs. Cooperation: A Benefit/Cost

Framework for Choosing Between Fully-Owned Investments and Cooperative Relationships," *Management International Review*, Spring 1988.

4. C. Oman, "New Forms of Investment in Developing Countries," working paper, OECD Development Center, 1984.

5. L. Franko, "New Forms of Investment in Developing Countries: Practices of U.S. Companies in the Automobile, Auto Parts, Food Processing, Pharmaceutical and Computer Industries," working paper, Center for International Competitive Analysis, Tufts University, 1985.

6. National Science Foundation, *Science Indicators, 1985* (Washington, DC: U.S. Government Printing Office, 1985).

7. H. H. Glismann, and E-J. Horn, "Comparative Invention Performance of Major Industrial Countries: Patterns and Explanations," *Management Science*, 34, no. 10 (October 1988), pp. 1169–87.

8. Edwin Mansfield, "The Speed and Cost of Industrial Innovations in Japan and the United States: External vs. Internal Technology," in *Management Science*, 34, no. 10 (October 1988), pp. 1157–68.

9. NSF, *Science Indicators*.

10. H. Fusfeld, and C. Haklisch, "Cooperative R&D for Competitors," *Harvard Business Review*, November 1985, pp. 60–76.

11. Hymer, *International Operations*.

12. R. Caves, "International Corporations: The Industrial Economics of Foreign Investment," *Economica*, February 1971, pp. 1–27.

13. S. Magee, "Information and the Multinational Corporation: An Appropriability Theory of Direct Foreign Investment." In J. Bhagwati, ed: *The New International Economic Order* (Cambridge, MA: MIT Press, 1977).

14. K. Harringan, *Strategies for Joint Ventures* (Lexington, MA: Lexington Books, 1985); and Contractor and Lorange, "Competition."

15. F. Contractor, "The Composition of Licensing Fees and Arrangements as a Function of Economic Development of Technology Recipient Nations," *Journal of International Business Studies*, Winter 1980, pp. 47–62.

16. R. Coase, "The Nature of the Firm," *Economica*, November 1937, pp. 386–405.

17. O. E. Williamson, *Markets and Hierarchies: Analysis and Antitrust Implications* (New York: Free Press, 1975).

18. J. F. Hennart, *A Theory of Multinational Enterprise* (Ann Arbor: University of Michigan Press, 1982).

19. E. Anderson, and H. Gatignon, "Modes of Foreign Entry: A Transaction Cost Analysis and Propositions," *Journal of International Business Studies*, Fall 1986, pp. 1–26.

20. J. Abegglen, "U.S.-Japanese Technological Exchange in Perspective, 1946–1981." In C. Uehara, ed.: *Technological Exchange: The U.S.-Japanese Experience* (New York: University Press, 1982).

21. L. Jacque, "The Changing Personality of U.S.-Japanese Joint Ventures: A Value-Added Chain Mapping Paradigm," working paper, Reginald H. Jones Center, The Wharton School, 1986.

22. Farok Contractor, *International Technology Licensing: Compensation, Costs, and Negotiations* (Lexington, MA: Lexington Books, 1981).

23. D. Ford, "Develop Your Technology Strategy," *Long Range Planning*, vol. 21, no. 5 (1988).

24. E. W. Eckard, "Altlernative Vertical Structures: The Case of the Japanese Auto Industry," *Business Economics*, October 1984, pp. 57–61.

25. D. J. Teece, "The Market for Know-how and the Efficient International Transfer of Technology," *Annuals of the American Academy of Political and Social Science*, 458 (November 1981), 81–96.

26. T. Ozawa, *Japan's Technological Challenge to the West, 1950–1974: Motivation and Accomplishments* (Cambridige, MA: MIT Press), 1974.

27. F. Contractor, *Licensing in International Strategy: A Guide for Planning and Negotiations* (Westport, CT: Quorum Books, 1985).

28. Harrigan, *Strategies*, and "Joint Ventures and Competitive Strategy," *Strategic Management Journal*, 9 (1988), 141–58; and P. Killing, *Strategies for Joint Venture Success* (New York: Praeger, 1983).

29. Harrigan, *Strategies*; and Contractor and Lorange, "Competition."

30. N. Adler, "A Typology of Management Studies Involving Culture," *Journal of International Business Studies*, Fall 1983, pp. 29–48.

31. G. Hofstede, "The Cultural Relativity of Organizational Practices and Theories," *Journal of International Business Studies*, Fall 1983, pp. 75–90.

32. J. Child, and A. Keiser, "A Development of Organizations Over Time." In P. Nystrom and W. Starbuck: *Handbook of Organizational Design* (New York: Oxford, 1918).

33. Anderson and Garignon, "Modes."

34. V. Balasubramanyam, *International Transfer of Technology to India* (New York: Praeger, 1973).

35. Ibid.

36. Y. Rhee, B. Ross-Larson, and G. Pursell, *Korea's Competitive Edge: Managing Entry into World Markets* (Washington, DC: John Hopkins University Press, 1984).

37. D. Encarnation, and L. Wells, "Evaluating Foreign Investment." In P. Grub, F. Ghadar, and D. Khambata: *The Multinational Enterprise in Transition*, 3rd ed. (Princeton, NJ: Darwin Press, 1986).

Chapter 4

Transferring Technology to China

*Jack N. Behrman, William A. Fischer, and
Dennis F. Simon*

TECHNOLOGY TRANSFERS

Though Chinese enterprises and government agencies (national, provincial, and municipal) are willing (even eager) to accept foreign technology, the willingness of the foreigner to release it and to incur the costs of transfer is often less strong. This is a result of a lack of information about China or a simple lack of interest. A Chinese radio factory interested in acquiring electronic memory operating systems (EMOS) semiconductor technology approached four major American complimentary metal oxide semiconductor (CMOS) manufacturers, but none was interested except at very high fees; the project seemed small to them. The Chinese domestic market was either closed to them, or unattractive, and there was concern that a future competitor might be generated. Rather than open up domestic markets to outside interests, China has sought technology through licenses and joint ventures for the explicit purpose of serving export markets. In one instance, a radio speaker enterprise spent $3.8 million for technical imports between 1983–87, resulting in exports of $1.5 million in 1987 and on anticipating level of $13 million in 1988.

China's efforts to limit the term of joint ventures have made some transnational corporations (TNCs) even less eager to do business there. This willingness is also reduced by a historical lack of appreciation in China (as in Taiwan) of the need to protect intellectual property, such as trade names, copyrights, and patents. Although China has enacted a patent law, increased competitiveness within the domestic Chinese market appears to be causing them to be more protective of intellectual property, whether or not there is a law.[1] Still, the development of enterprise that arises through invention and innovation that leads to profit for the principal actors neither permeates China's economic history nor is a major vehicle at present.

Case A: Cotton Textile Factory

Approximately 20 percent of the output of this cotton textile factory[2] was produced using imported technology, embodied partly in foreign machinery from Europe and Japan. The gap between the factory and world technological standards was not large until the mid-1970s, though prior to that time there was little interaction between the factory and foreign equipment manufacturers. Consequently, it had little knowledge of new technologies, and there was little diffusion of information among textile factories.

With the modernization policies in 1978, this factory became a leader in importation of foreign technology, looking to overseas markets. If the products were competitive enough to export, the factory received a higher price and the workers a larger bonus.

The criteria for selection of the source of new technology were: (1) price, and (2) appropriateness (that is, could China use it? Could it be adapted to factor conditions? Could it produce good quality products?) In this case, the answers were affirmative, and the enterprise was able to export the resulting product. But the enterprise managers concluded in retrospect that they did not need to import so much foreign equipment. They did so not understanding the international market and sought to be a "model" for others. It did succeed in this, and assisted others, but the cost was high.

The willingness of the owner of proprietary technology to engage in transfers is also related to the terms of the agreement and the pursuit of common adjectives.

A reluctance to transfer technology has arisen at times from disagreements over policies of a joint venture—such as markets to be entered and rates of expansion. In the case of the joint venture between Parker-Hannifan (Ohio) and the Hubei Automotive Industry Corp., considerable technology was transferred to the venture to rectify the poor state of seal technology in China. Several Chinese engineers were sent to P-H labs to learn quality control and high precision mold design and use. Disagreement arose over markets, with Hubei insisting on expanding exports promptly, especially through Parker-Hannifan; whereas P-H sought sales in China. Hubei also wanted to expand production and open a new plant to serve a dynamic auto industry in China. P-H considered this growth much too fast and was fearful of being pressed itself to take an oversupply of the Chinese product. A fast production pace at Hubei damaged some of the tools and equipment, reducing quality. The Chinese considered that P-H's disagreement on policy was reflected in a loss of interest in the venture and inadequate support of production. After much tension, the problems were somewhat relieved through compromises on each side.[3]

There are some 30 provisions in a usual licensing agreement over which disagreements can arise and thereby alter the eagerness of the license. For example, the right of the licensor to receive back any improvements made by the licensee would tend to reduce the royalties requested. Similarly, extension to the licensee of an exclusive right to use the technology would increase the royalty rate. As with most developing countries, China has continued to undervalue technology (seeking to negotiate low royalties) as compared to world practices or what a potential licensor considers appropriate. Thus under a license from an American producer of generator technology, the licensor receives royalties on all exports by the Chinese enterprise but only the contracted transfer fee for any level of domestic sales. This reflects a "zero-sum" mentality in which the Chinese side considers that anything given up to the other party must represent an absolute loss to itself. This attitude perhaps is the result of China historically being principally a trading society. A tape recorder factory that bought production equipment and technology from Japanese sources has found its line down frequently for lack of Japanese components. The delays were attributed by the licensee to Japanese favoritism to other customers rather than the equally likely inadequacies of the Chinese transportation system.

Case B: Central Transformer Factory

A factory producing a variety of electrical and electronic products determined that it needed foreign technology—mainly in the form of new equipment for production—in order to: (1) upgrade its technical capability, (2) gain a competitive edge in the Chinese market, (3) improve export ability, and (4) achieve a tie-in with the technology supplier so as to use it as an export channel.

The existing processes were largely manual, whereas more modern machinery was automatic and computer-controlled. The factory management was aware of developments abroad but could not afford new equipment. When it was informed (through government channels) that used equipment was available in the United States, it moved quickly to acquire 56 old machines and 24 new ones needed to complete the lines; the latter coming from Poland, Holland, Switzerland, and the United States. The factory was to pay for some purchases from the American supplier through exports of the final product, to be sold by the latter, thereby establishing an export channel for later Chinese production.

Before the purchase of machinery, the factory produced 40,000 pieces per year, with a quality rate of between 40 and 60 percent acceptable. The Chinese management thought that the comparable U.S. output of the equipment was 150,000 pieces a year, with 70 to 80 percent acceptability. Once the machines were in place, the managers recognized that several technical problems would have to be overcome. But accompanying documentation was not sufficient to allow operation of the equipment, and much of it remained idle.

Because of the lack of understanding by the transferrer, and by the transferor of the operating environment and the accompanying technology needed, the factory's ability to export and generate the foreign exchange expanded by the technology acquisition was severely limited. Without the additional technical know-how, the transfer was approaching a disaster.

The factory managers provided the following comparison of national suppliers of technology. *Germans* were seen as the most serious, and diligent but least easy to negotiate with, being resistant to bargaining on prices. *Japanese* were seen as ready to come to an agreement, being more flexible on prices ("starting high and cutting easily"), but reluctant to transfer "real technology" (meaning the most advanced); they provided excellent service until final completion of the transfer. *Americans* were seen as ready to bargain, making concessions even after agreements were reached, and willing to share existing technology; however, they left the job when their time was up. "They are very careful of their own time and don't want to stay in China too long."

Each of these attitudes affected the willingness of the supplier to support the technology and equipment transfer with full "working know-how." In addition, without adequate knowledge of what it "needs to know," the Chinese factory often obtained less than was necessary to make the transfer successful.

Of even greater importance to the effectiveness of technology transfers is the ability of the licensor to transfer the technology so that it can be readily understood and put into practice by the licensee. Not all potential licensor are so capable. TNCs with extensive practice in technology transfers are more likely to know how to do so effectively, but the transferred technology may be so standardized both in design and in transfer approach that the transferor cannot make appropriate adaptations for different cultural or economic/social settings.[4] Some TNCs essentially transfer a standardized package and let the licensee sort out how it is used. Other companies, who are new to the game, learn by doing and therefore are not to adapt it to different conditions and different materials. The radio factory mentioned above turned to a smaller U.S. company that agreed to supply know-how and guide the purchase of appropriate equipment to original equipment manufacturer (OEM) companies in the United States. But the licensor had little previous experience with technology transfer and did not know whether it could successfully transfer all software or could get permission from the U.S. Department of Commerce to do so. Preliminary trips at a presumably high cost to the licensor and a Commerce official were required to provide assurances of potential successful transfers.

Not the least of the problems facing licensors is their unfamiliarity with different cultures and the way in which they are related to technology. Without such background knowledge, they cannot anticipate the problems likely to be encountered and they will sometimes underestimate the costs of transfer, only to become frustrated later as costs rise in meeting unanticipated situations. Accordingly, the desire on the part of many developing countries, including China, for technology from medium or even small-size companies frequently gains a license agreement, but not necessarily an effective transfer, because of a lack of appropriate experience.

Given time and a willingness on the part of both parties to go through the process of learning, most licensor eventually find a way of transferring technology effectively. Some, however, simply give up the game, being unwilling to take the time or make the effort to do so.

Case C: Northeastern Office Equipment Enterprise

The enterprise produced six different office copiers and controlled 47 percent of the Chinese market, selling 30,000 units annually. It anticipated producing 50,000 and exporting 60 percent of this to Southeast Asia, earning some $2 million in foreign exchange.

The technology transfer was from Japan to help produce photo receptor drums, using Japanese equipment, at a cost of $4 million for technology and know-how. Of the 217 pieces of equipment used in production, 184 were imported along with the technology.

An underlying objective of this technology acquisition was to upgrade domestic production quality so as to be able to export. Some exports were, in fact, made to the Japanese licensor. The Japanese company was selected because its copier was suitable to production by the Chinese enterprise, the licensor was strong in copier technology, its delivery schedule was 10 months earlier than the others, and a preferential price was given for technology since three production lines were being ordered also.

Some of the required technology had been shared with another licensee of the same Japanese company, and suppliers in China had been assisted. The indigenization of supplies and technology had also been assisted by the Municipal Economic Commission who helped find an R&D institute that could design components and subassemblies not included in the transferred Japanese technology. Even so, not enough attention had been paid to the problems of indigenization of the toner and photo receptor technology when first imported. In the future, managers concluded, they should use more Chinese raw materials. The activity was hampered also by the inability to talk directly to suppliers of the Japanese OEM licensor. The Chinese management felt that it lacked sufficient blueprints and technical materials to master the technology on its own. The Chinese learned a lesson not to let such an inadequacy arise again. The Chinese factory continued without complete information on maintenance. Sometimes replacement parts for machinery were difficult to obtain—again because the Chinese could not deal directly with subcontractors to the Japanese licensor, which also handled all equipment sales in the agreement.

TECHNOLOGY ABSORPTION

The most important aspect of successful technology transfer, however, is the *ability* of the host country and licensee to absorb the technology effectively. This ability is taken for granted on the part of most developing countries. As a consequence, not enough is done to make certain that the receptiveness actually exists.[5] This is frequently the case in China, and the inability particularly to absorb *sophisticated* technology is not always recognized by the Chinese recipient. In the case of Shanghai-Foxboro, the Chinese wanted a more sophisticated technology, before they had mastered basic analog technology in its process control devices; Foxboro refused to acquiesce.[6]

A substantial effort is required in building a scientific and technological infrastructure not only within the recipient enterprise but in the institutional environment surrounding and supporting it. The recipient should be adequately prepared technically (as well as managerially and psychologically) to accept, adapt, and use the technology. And suppliers and vendors should learn to raise their own abilities to match the imported technology so as not to undercut the value of the technology received from abroad, by producing components of a quality lower than that needed by the manufacturer of the filled product or not knowing how to maintain and repair the product. Such gaps will lead to breakdowns with the equipment in disuse longer than is appropriate, leading to high costs and low returns for the technology transfer.

The licensee itself must be technically, psychologically, and organizationally prepared to use the technology, or else it simply sits on the shelf or is applied ineffectively.[7] An enterprise in northeast China was urged by the electronics industry to acquire semiconductor technology. Consequently, it bought a used assembly line from a U.S. company producing CMOS chips, but did not acquire the "know-how." It could not produce CMOS chips with it so it reverted to its usual production of bipolar 10-micron chips, of a 20-year-old vintage. It has been unable to enhance the "technique" to use the line for CMOS or any improved chips. The reasons for technology import problems are manifold. A commentary in Shanghai's *Liberation Daily* points out three major problems. (1) lack of industrial branch development plans to guide technology imports, with the result that emphasis is placed on short-term economic benefits to the neglect of long-term considerations and as-similation, (2) ineffective functioning of the technology import organizations and technology assimilation bodies, with a lack of coordination and unified command by an authoritative body, and (3) a shortage of funds to assist in assimilation, with the result that many key technology projects remain unfinished.

In addition, there is often a lack of understanding of the basic manufacturing techniques, limited competence of engineers and technicians, maintenance problems, and acquisition of equipment that requires too high a degree of sophistication.

The late Frank Bradbury, one of the pioneer students of technology transfer, observed that "a characteristic of technology transfer is that some change must occur: either in the technology transferred or the recipient organization."[8] Despite a willingness to receive technology, however, there is frequently an unwillingness to make the necessary adaptations or changes in the enterprise's organization, power structure, production layouts, employee skills and level of employment—all of which upset the existing system. Relatively few developing countries have created the scientific and technological infrastructure necessary to receive, adapt, and use technology transferred from abroad.[9] To conserve the resources of the host country and make certain that the development occurs in the most effective manner, it is necessary to have a scientific community that can assess "appropriate" technology and can also distinguish between what is necessary for the national interest as distinct from what is desirable from the interests of the receiving enterprise.

These distinctions are necessary if an approval process is to be useful at all. The Chinese government, for example, seeks import substitution as a means of generating economic growth, but the protection of the domestic market, characterized by large amounts of unsatisfied demand, provides little incentive for a Chinese enterprise to apply technological changes either to expand the market or to reduce costs.[10]

Thus there is little linkage between the development strategy of the government and the interests of enterprises. Only if there is competitive pressure—generated either by additional enterprises in the local market or an opening to the international market through imports or exports—does the pressure increase to make effective adaptations and use of imported technology.

Case D: Eastern Electric Motor Factory

This factory began research into submersible pump technology for oil recovery in 1975 and went into production in 1979. Subsequently, it had produced 910 pumps for use in Chinese oil drills in more than 10 different models of its pumps. However, they did not meet quality expectations and could not be produced in enough volume to meet domestic demand. Complaints about the product had come in from the fields.

To remedy this situation, the factory bought technology from an American firm plus equipment for $8.5 million. It required three years (1984–87) to begin production of the new models. American-produced pumps were imported in the meantime. The technology transfer involved over 8,500 documents, including some 870 technical specifications and drawings. The factory mobilized 100 engineers to translate and digest this material. The licensor trained the Chinese engineers in the United States and provided 20-man months of technical assistance on site in China. To digest the information, an additional $1.5 million was spent for sophisticated machine tools; additional tools were produced by the Chinese enterprise to American designs.

The enterprise anticipated having all technology absorbed in five years. It had established a technology information office to conduct visits, develop engineering specifications, data collection, training, and indigenization of the imported technology so Chinese suppliers could be substituted eventually.[11] The 20 engineers trained in the United States became trainers in China of the remaining staff, the best people being chosen for these positions.

The process of absorption was guided by two commandments: (1) change nothing, and (2) pay attention to quality. To be sure and learn all the licensor had to offer, the enterprise wished to follow the specifications and directions explicitly. A major problem arose relative to the second guideline; the Chinese had their own QC system and the American system was not considered "appropriate." The American licensor required every worker to "be very alert," but this was not possible in China, according to the manager of the Chinese enterprise. American workers were seen as well trained and able to make decisions, but, said the manager, "Chinese workers are not so capable of participation. We figure it will take two more years to reach the level for them to be able to participate in the way assumed by American quality control systems." Japanese quality systems were studied but not wholly adopted either. The manager observed, "The Japanese give their workers drawings and expect them to live up to these drawings. We have been trying this for four years, but some of our workers falsify their reports. They don't do their job well."

DIFFUSION OF TECHNOLOGY

To make the import of technology most effective in supporting economic growth, it is necessary for acquired technologies to be diffused throughout the society. Lead sectors should be stimulated to employ advanced technologies, which can then impress both suppliers and vendors with the necessity to raise their technical abilities. In turn, these secondary and tertiary industries will require scientific and technological support from the national industrial research network. If there is no such network, the imported technology is likely to be less well suited for the nation's needs, though it is seen as desirable by the enterprise itself. In China, there are some formal networks for diffusion through the ministries of various industries, but there are few R&D institutes attached to them that are also knowledgeable about what is needed by enterprises. A radio factory was assisted in the design of CMOS semiconductors by several research institutes under the Chinese Academy of Services; these institutes helped with mass-consumer products, whereas custom-designed products were generated by the factory itself— just the reverse of what might typically be expected in the West.

In some instances, at the initiative of the licensees, technology will be shared with other Chinese enterprises—unless prohibited in the agreement. In one case, the Chinese licensee was able to adapt an automated Japanese production line to less-automated function. It then passed this adaptation on to other Chinese factories. Diffusion of the imported technology is often thwarted by the inability of supplier companies in China to absorb the technology appropriate for the components they are to produce. The efforts of the Shanghai-Foxboro joint venture to increase local value-added, from 10–20 percent up to 30–40 percent, were frustrated because suppliers could not produce high quality sophisticated electronic components, could not handle the special metals, and at times could not find appropriate material inputs. For technology transfers to make their maximum contribution to China's development, an independent scientific and technological community is necessary. The Chinese community was systematically destroyed in the cultural revolution, but even this group was not oriented toward science for application and was hardly independent. Support of applied science must be provided in enterprises by both the government and educational institutions. To some extent this is being done by government policies removing the Academy of Science research institutes from grants-type funding and into a market-driven, fees-for-success approach. This is a major departure from tradition, however, and it will take time to become established.

NEGOTIATIONS

Only if there is sufficient scientific personnel and technological expertise available to assess imported technology will the country be able to determine the most appropriate technology for the country's needs and to bargain for it effectively. It must have sufficient ability to determine some of the needs beforehand, not relying wholly on the foreigner to suggest them. China's present orientation to technology transfers through joint ventures is to wait for the foreigner to take initiatives, which is basically a function of the fact that China does not have sufficient information about what is available, nor who is capable and interested. The Chinese do make some prior technology selections through pilot licensing agreements and through searching the literature and asking Chinese companies to identify acceptable foreigner suppliers of similar products. For example, a textile machinery enterprise (a dominant producer in the Chinese market) turned to Chinese customers of European manufacturers to identify the best foreign technology supplies. The subsequent examination of potential suppliers took three years, however, because of a basic lack of continuing information and contacts. The enterprise was forced to learn all it could from magazines, brochures, reports, and then visit end-users in Europe, before narrowing the field. The final supplier decision was based on price—as is usual in Chinese negotiations—because several potential licensor have similarly advanced technologies.

Negotiation of joint ventures are more complex, since the contribution of the foreign TNC is more than technology. More information is needed, and this is difficult for the Chinese to obtain. This lack of information limits the Chinese bargaining position and makes them wary in the process, simply because they do not have an adequate appreciation of the universe with which they are dealing. China feels itself weak compared to the strengths of the foreigner and the information the foreigner might have; therefore, they often bargain from an extreme and rigid position, which makes it difficult to come to an appropriate agreement.

Without an adequate scientific and technological network, it is virtually impossible to plan for international technical cooperation or to make certain that the resources of the country are used effectively. The foreigner, therefore, faces "on-again, off-again" behavior, resulting in a high frustration level, an erosion of "goodwill," and a reduction in the contribution they are willing to make.

Case E: Tanggu Chemical Factory

During 1972–74, the State Planning Commission, the Ministry of Chemicals, the Municipal Chemicals Bureau, and the Technology Import Corporation decided among themselves to expand the output of petrochemicals at the Tanggu Chemical Factory (TCF) in northeastern China. A modern processing facility would be imported from abroad, along with the requisite technology for production of an intermediate chemical and maintenance of the equipment. A German company was finally selected as the source. TCF was the supplier of over half of the existing Chinese domestic market for the chemical it produced.

TCF had some of the best petrochemical technicians and technical experts in its management and had examined foreign technologies on occasion. It was also an adviser to other petrochemical plants and to the bureaus responsible for this sector on technical matters and machinery imports. Negotiations were conducted between the German company and the Technology Import Corporation though various other agencies, and TCF representatives also participated.

In an effort to save foreign exhange and as a direct result of the intervention of Ministry of Finance officials, the factory "unbundled" the turnkey package offered by the Germans and insisted on providing the water treatment facilities from domestic sources. This was a serious mistake as the Chinese side did not fully appreciate the rationale behind the processes and standards for incoming water, nor could domestic manufacturers satisfy these requirements. The result could have been disastrous. As it was, because of corrosion from water impurities, all heat exchanges had to be shut down, and the factory had run for only four months before requiring major repairs.

It required three months to replace the damaged equipment and to institute the suggestions of the licensor, such as putting additives in the water, upgrading the water treatment facility, and raising turbidity standards. Over the course of 14 months of operation, a variety of problems shut down the plant 33 times. The total cost of the equipment replaced and the repair work amounted to $4.2 million—compared to the $200,000 saved by unbundling the piping requirements, and an unreported amount spent on the water-treatment facility. To this must be added the cost of delays. The remedies of the licensor served to solve the problems and from mid-July 1979 to mid-1980, the factory averaged 90 percent of designed capacity.

The basic problems stemmed from (1) the assumption by the Chinese that they understood international standards, when they did not, and (2) a lack of expertise within the local science and technology community to adopt and adapt the incoming technology and find solutions for breakdowns.

There is also too great a distance between the officials in enterprises and those in government for dealing with scientific and technological matters. The former remain uninformed, whereas the views of the latter remain largely irrelevant to industrial needs—again leading to difficult or fruitless negotiations. A continuing link between government R&D institutes and policymakers and those who are the actual recipients of technology is required. This structure does not exist in China save in a few instances. In many sectors, such collaboration is entirely lacking. Furthermore, although steps are being taken to remedy these problems, during the process of rectification a good many false starts and detours can be expected. Without correction, however, unsound or irrelevant technologies will continue to be selected; attempts at adaptation will be fruitless; no adaptation will occur when necessary; and ineffective assimilation or use of imported technology will continue.

The ultimate contribution of imported technology, therefore, will continue to be minimized rather than maximized. Without such a network, the government cannot ascertain needs, assist enterprises, nor exploit their latent capabilities so as to make effective use of foreign technology. Such a network would help to examine the impacts of the technology and the opportunities it opens up. Technology is not merely an addition to existing resources. It can, and does, change labor/capital ratios, the amount of labor required, the skills required, and the material inputs required. It also changes the necessity for repair and maintenance, and even the patterns of use of the final product.

Consequently, it is necessary to have a network that can help in the selection of: (1) the particular technologies to be transferrred, (2) the enterprises that can, and will, make effective use of, (3) the means for making appropriate adaptations to the technology given domestic demands and material supply, and (4) the technologies necessary to stimulate export abilities and, therefore, constituting an engine for economic growth.

China recognizes the *need* to receive technology, and it is *willing* to receive it. It has the ability to search out *willing* suppliers. But Chinese enterprises are less capable of assessing the *ability* of suppliers to transfer the technology effectively. And enterprises overestimate their own *ability* to absorb and adapt foreign technologies, especially the more sophisticated.

Notes to Chapter 4

1. See William A. Fischer, and Dennis F. Simon, "The Acquisition of Foreign Technology by Chinese Enterprises." In *Technological Change & Economic Performance in the People's Republic of China* (Washington, DC: Woodrow Wilson Center, forthcoming).

2. This, and the following case illustrations are drawn from Dennis F. Simon, and William A. Fischer, *Technology Transfer to China*, (Cambridge, MA: Ballinger, 1989).

3. *Business China*, May 16, 1988, May 30, 1988, June 20, 1988.

4. David J. Teece, *The Multinational Corporation and the Resource Cost of International Technology Transfer*, (Cambridge, MA Balliner, 1976); and William H. Davidson, *Experience Effects in International Investment and Technology Transfers*, (Ann Arbor: U.M.I. Research Press, 1980).

5. Richard D. Robinson, *The International Transfer of Technology*, (Cambridge, MA: Ballinger, 1988), see especially Chapter 6.

6. *Business China*, June 27, 1988.

7. That this can occur even in technically sophisticated environments such as the United States is seen by the persistence of American machine tools users to employ expensive "flexible" machinery in exactly the same fashion as they used the traditional technologies it replaced. See Ramachandran Juikernow, "Post Industrial Manufacturing," *Harvard Business Review*, November-December 1986, pp. 69–76.

8. Frank R. Bradbury, "Review of Technology Transfer Studies." In K. M. Setron, H. Davidson, and J. Goldbar, eds.: *Industrial Application of Technology Transfer*, (Leydon, Netherlands; Nordhoff, 1977).

9. See Jack N. Behrman and N. W. Wallender, *Transfer of Manufacturing Technology within Multinational Enterprises*, (Boston: Ballinger, 1976); and Jack N. Behrman and W. A. Fischer, *Overseas R&D Activities of Transnational Companies*, (Cambridge, MA: Oelgeschlager, Gunn, and Hain, 1980).

10. This is referred to as "big market mentality" in William A. Fischer, *Chinese Manufacturing Capabilities*, working paper, Center for Manufacturing Excellence, Graduate School of Business Administration, University of North Carolina at Chapel Hill, March 1989.

11. Technology information offices are a recent organizational innovation, created by Chinese enterprises to facilitate the acquisitions and implementation of foreign technology. See Simon and Fischer, *Technology Transfer*.

Chapter 5

Transfer of Technology through Backward Linkages of FDI—A Policy, Programs, and Institutions Approach

Joseph Y. Battat

Among the major benefits expected from foreign direct investment (FDI) in a host developing country are the generation of additional economic activities, the development of entrepreneurship, and the transfer and diffusion of technology and management know-how. The extent and nature of those benefits are functions of the degree and character of the integration of FDI within the local economy. The more the integration or linkage and the more it matches the country's development needs, the more benefits are generated within the domestic economy.

This chapter focuses on the policy, programs, and institutions used to foster backward linkages between foreign affiliates and domestic firms in four countries. It briefly presents definitions and concepts related to backward linkage. It summarizes the experiences of South Korea, Taiwan, Singapore, and Ireland in their promotion of backward linkages with an emphasis on the mechanisms and approaches being used to that end. Finally, it draws lessons to be learned from those experiences.

CONCEPTS AND NOTES

Backward linkages stimulate activities in the domestic economy upstream from FDI operations, e.g., supply by domestic firms to foreign affiliates of raw materials, parts, components, and services. Other benefits from indirect linkages with FDI, including the transfer and diffusion of technology through human resources mobility and the development of entrepreneurship and infrastructure, are not examined.

We start from the premise that a backward linkage of FDI into the domestic economy offers a powerful mechanism for the transfer of technology. Foreign affiliates contracting with domestic vendors for the supply of metal or plastic parts and components, metal working, printing, or packaging, which provides the opportunity to work with their vendors on product development, is one of the many facets of technology transfer,[1] as are the identification, transfer, and use of needed technology and assistance in enhancing management production and operations system. To effect the transfer of technology through backward linkages, many governments have taken measures to foster such linkages. The chapter reports the results of a four-country survey of backward linkage programs, using an environmental approach, i.e., policy, programs, and institutions.

Backward linkages programs may rest on the imposition of local content requirements, whether economically justifiable, and the prohibition of the importation of parts and components, coupled with a protectionist import substitution policy. This *administrative-oriented* approach usually leads to inefficiencies and technological stagnation, particularly when it lacks the mechanisms that eventually expose manufacturers to domestic and particularly international competition, e.g., imposing export requirements, or the gradual removal of tariff protection. Moreover, there is often no stipulation as to whether local content requirements lead to backward linkages into domestic firms because local production may occur within the foreign affiliate itself, in other foreign affiliates, or in domestic firms. Whereas they offer benefits to specific and narrow groups and organizations, local content requirements implemented under an inappropriately protectionist regime lead to economic distortions, and eventually to diseconomies and erosion of much of the sought-after benefits.

Mexico's backward linkage policy typifies this administrative-oriented approach. Introduced in 1973, the policy produced highly inefficient and entirely locally oriented domestic industries. Likewise, the operations of many of its foreign affiliates became inefficient and exclusively geared to the protected local market. The policy led to economic distortions and the erosion of much of the sought-after benefits expected from FDI and backward linkages, to the point where the national economy could not sustain it. In 1987, Mexico began a liberalization process, including the reduction of trade barriers and removal of local content requirements.

Rather than counting on legal or negotiated requirements, efforts at fostering backward linkages in a few countries have relied on policies and promotional measures to enhance the domestic business environment as well as the capabilities of domestic firms within a market framework. These measures are designed to remedy imperfections in market mechanisms and enable governments to play the role of broker, matchmaker, and facilitator between foreign affiliate and potential subcontractors. The philosophy of this approach, labeled *market-oriented*, remains one of freedom of decision, whereby both domestic and foreign firms are free to take advantage of the policies and measures offered and to enter into business arrangements.

Measures to encourage market-oriented backward linkages with domestic firms are economically less distorting, often politically more attractive, yet more complex and difficult to implement. Unlike local content requirements, this approach has demanded from governments deliberate and widespread efforts at enhancing domestic business environment and firms' capabilities to render them more attractive vendors to foreign as well as large domestic firms. In market economies, this approach renders domestic vendors increasingly internationally competitive, because they have to become attractive enough to induce multinational companies (MNCs), which are exposed to the rigors of the international market, to procure domestically. Moreover, it is the domestic firms that retain the benefits of the additional localized production, including revenues, and the development and diffusion of technical, managerial, and marketing competence, developing in the process a comparative advantage within the sectors targeted for linkages.

A policy of encouraging market-oriented backward linkages requires a high degree of sophistication in terms of policy formulation and implementation. Figure 5.1 illustrates that these policies must be coordinated with a wide range of other economic strategies and policies:

- A backward linkage program must be consistent with the national economic development strategy.
- Linkages are more likely to happen in sectors for which a country has developed, or is in a position to develop, a certain capability and eventually a comparative advantage. Thus industrial policies and backward linkages are closely interrelated.
- In almost all cases, efforts at enhancing backward integration lead to the strengthening of small and medium-size enterprises (SMEs) to position them to enter into subcontracting relationships with foreign affiliates.
- FDI and trade policies affect backward linkages inasmuch as the nature of the business of foreign affiliates and entry requirements imposed on them help determine linkage opportunities. Import policies for materials, semifinished and finished products, and capital goods have direct effect on the ability of domestic firms to become vendors to foreign affiliates.

Backward linkage efforts must be supported by a number of other policy measures:

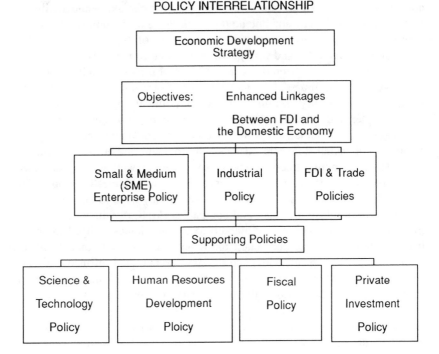

Figure 5.1. Backward linkages of foreign direct investment into the domestic economy.

- Science and technology policy must provide technological support to enterprises for product design and development, and improved productivity and production facilities.
- Human resource development measures are needed to train domestic managerial and technical personnel to position enterprises as potential suppliers to FDI firms.
- Fiscal policies may create or remove tax-based impediments to subcontracting. Also, tax incentives are useful to influence foreign and domestic firms' behavior for backward linkages.
- Private investment policy, including financing schemes for local industry, may help remedy difficulties faced by small and medium enterprises in accessing capital markets.

A survey of backward linkages of South Korea, Taiwan, Singapore, and Ireland was conducted in the fall of 1989 in the form of interviews of government and paragovernment agencies, a sample of foreign affiliates and domestic firms operating in these countries, and business and industry associations. The survey focused on understanding the policy, programs, and institutions established to foster backward linkages of foreign affiliates with domestic firms. Using the policy framework (Fig. 5.1) as a guide and with the above comment in mind, the experience of the four countries—South Korea, Taiwan, Singapore, and Ireland—in fostering backward linkages are synthesized below for two purposes: to improve understanding of how backward linkages may be fostered, and to draw lessons from their experience.

The survey led us to note that all programs are closely related, though differently, to existing SME (small and medium-size enterprise) programs. In South Korea and Taiwan, the backward linkage program is subsumed under the SME program. In Singapore and Ireland, the backward linkage program stands alone and is distinct from, though closely related to, the SME program.

This observation merits two comments. First, although domestic firms involved in backward linkages may be of all sizes, government efforts seem to be centered on domestic SMEs, for perhaps two reasons. For one, as a category of business, they are perceived as the most feasible candidates for backward linkages, because they form the foundation of the national industrial structure and in the manufacturing sector, their businesses tend to be in supporting and parts and components industries. Also, unlike larger domestic firms, they lack the financial, technical, and managerial capabilities to become vendors to foreign firms, and they need government assistance in these areas. For the second reason, not enough information is available to explain why backward linkage programs are established separate from or as part of SME programs. We note, however, that in both Singapore and Ireland, where the backward linkage programs are distinct from the SME programs, FDI has played a central and prevalent role in their economic development, and their foreign economic policies are truly outward-oriented.

Another point the survey has shown is that, although the programs target backward linkages of foreign firms with their domestic counterparts, they also actively seek to develop linkages between SMEs and larger domestic firms. This is particularly so in South Korea, where large industrial conglomerates, the chaebols, dominated the economy. In market-driven economies, the programs and measures designed to make SMEs able to supply foreign affiliates are similar to those needed to make them supply large domestic firms, which themselves face internal as well as international competition, and are often MNCs in their own right.[2]

POLICY

The overall objective of a policy of backward linkages is in itself simple: it is to locate upstream economic activities in domestic firms—usually SMEs—

rather than in foreign firms whether inside or outside the country. It is in its design, structure, and implementation that the policy has shown much of its complexity, as the backward linkage policy framework (Fig. 5.1) would suggest.

In all surveyed countries, the backward linkage policy has addressed several issues.

Underlying Economic Philosophy

Policymakers are soon faced with two philosophically distinct approaches to backward linkages, the administrative- and the market-oriented approaches. The administrative-oriented approach, such as the one Mexico had adopted, is very attractive politically, for at least two reasons. Of the two approaches, it requires the least amount of effort at formulating and implementing the policy, i.e., passing the legislation and related regulations and managing their enforcement. Also, it creates over a short period of time a whole population of domestic vendors, because they enjoy a captive domestic market. The experience of countries that have adopted this approach shows that, unless the policy contains built-in mechanisms to promote efficiency and technical innovation, it has led to inefficiencies, economic distortions, and the loss of most of the expected benefits. Like Mexico, many of the countries that had adopted this approach have liberalized their economic regulations to foster a more market-oriented environment.

The premise of the market-oriented approach is that when a favorable business environment is created and domestic firms' capabilities are enhanced, backward linkages become attractive options to foreign firms and will take place on the foundation of economic rationale. Compared to the administration policies, this approach is clearly more elaborate and requires more effort to formulate and implement successfully. It gives less control to the authorities in its implementation, because after all, linkages will be decided principally by the market. Yet, it has been found that the direct intervention of the authorities to remedy imperfections in market information has been necessary. The authorities in all four countries work directly with foreign and domestic firms to identify backward linkage opportunities and act as brokers and matchmakers between potential buyers and sellers, as discussed later in the Programs section.

Taiwan, Singapore, and Ireland have adopted essentially the market-oriented approach. South Korea has a hybrid philosophical approach to backward linkages, although it leans more toward the market-oriented. Its backward linkage policy contains two administrative-oriented features. Each year the government establishes a list of "priority" parts and components that businesses must buy rather than produce internally. There is another list of "procurements" products that government agencies must purchase from SMEs up to a predetermined percentage of their needs. Also, as part of the objective of increasing localization and developing Korea's weak parts and components industries, the authorities use strong pressure on large companies to establish domestic vendor networks. In view of the government's close working relationships with and leverage over the very large firms (chaebols), such pressure is tantamount to legislation and has effectively elicited the chaebols' positive response.

Broadbase Political Commitment to Policy

Market-oriented backward linkages and their supporting programs take a long time to develop. Firms, foreign and domestic, need to perceive their vendor/buyer relationships as potentially long term and stable before making the strategic business decision for local subcontracting. For that to happen, a broadbase understanding and sharing of the objectives and policy of backward linkages must be gained, and long-term, sustained, consistent, and high level political support and commitment for allocation of resources must be present.

In all four countries, it was striking to note the presence of such a political support and broadbase understanding and sharing of backward linkage policies' objectives. And in all four, budget allocations for the SME and linkage programs have increased substantially. Notably, South Korea's support for its SME and backward linkage policies comes from the president himself; he chairs semiannual national conferences on SMEs, and his monthly visits to an SME are regularly covered by the media. More importantly, South Korea's authorities have provided generous and sustained financial and other support, including, beginning in 1990, an increase in the budget allocation of 1.3 trillion won for a five-year period. In Singapore and Ireland, the backward linkage programs have attracted a dedicated, high caliber, and result-oriented staff, who exhibit a clear and shared understanding of policies.

Consistency with National Economic Environment, Development Strategy, and Policies

An effective backward linkage policy must be based on national economic strategy, be consistent with economic objectives, and closely associated with other economic policies, such as FDI, trade, industrial, and SME policies. In Taiwan, Singapore, and Ireland, where FDI plays a key role in national economic development, backward linkages are considered an important source not only of added economic activities, but also for the transfer and diffusion of technology and management know-how through foreign affiliates. Those countries' backward linkage programs are clearly geared to linking foreign affiliates with domestic vendors. By contrast, FDI plays a less central role in Korea's economic development, and linkages are enhanced more between SMEs and the chaebols than with foreign firms. Also, Korea's linkages are closely geared to its import substitution policy and support the aggressive development of its previously neglected SMEs.

In all four countries, the linkage programs have access to fiscal, investment, science and technology, business development and human resource development support programs, without which their backward linkage policies would not have been viable. So, it is essential that a backward linkage policy be founded on a host of strong supporting policies and programs, as we see in the next section.

Targeted Economic and Industrial Sectors

In all four countries, the linkage policies target both the sectors of potential buyers and those of potential vendors. On the buyer side, linkages are fostered in specific sectors as a function of where the highest potential demands for linkages exist, e.g., sectors in which FDI presence is high, such as electronics and food and drinks in Ireland, and information technology in Singapore, or as a function of national industrial policy and priority, as in the case of biotechnology in Singapore.

On the vendor side, linkages are fostered in those sectors in which the country possesses a comparative advantage and/or adequate production capability, typically metal working, light engineering, plastic parts and components, and mold and die manufacturing, which are targeted for development, such as information technology in Singapore, and which enjoy natural protection, e.g., supporting industries such as accounting and legal services, packaging, printing, and food and fluid handling systems.

Targeted Firms

Within targeted sectors, there is further need to target firms to participate in the linkage program. Within the buyer sectors, some backward linkage programs (Ireland and Singapore) target specific foreign firms using as selection criteria the firms' technology, reputation, volume of business, attitude to the linkage program, and potential local added value, to name a few.

In all four countries, targeted potential vendors belong to the category of SMEs for the simple reasons that they form by far the overall majority of firms, present the highest potential for development and expansion, yet need all kinds of support to bring them up to the level where linkages are feasible.

Further targeting among SMEs is needed, because there are too many of them chasing limited public resources. Using a "winner strategy" approach, the authorities have tended to support those SMEs that have the highest probability of success as vendors. The criteria of selection include the presence within a SME of a high quality management with a developmental mindset, able to adopt modern management systems to meet the professional rigors imposed by MNCs on their vendors. Its products have to be in demand by the MNC affiliates. It must exhibit capabilities for quality products and services, reliability, delivery, and cost competitiveness. And its management must be receptive to and be willing to develop a close and cooperative working relationship with the linkage program.

Linkage Program as a Stand Alone or Subsumed under the SME Program

Policymakers must decide what relationship the linkage policy should have to SME programs. It was noted earlier that all linkage policies have targeted SMEs as the type of economic organizations suitable for potential linkages and that they rely heavily on the assistance and support programs established for their SME policies. It is not surprising then, as mentioned in the previous chapter, that South Korea's linkage program is subsumed under its SME program. In contrast, with quite outward-oriented FDI policies, Ireland and Singapore have stand-alone linkage programs that, however, are closely associated with their SME programs. Taiwan's situation is similar to that of Korea, with the presence of its Center-Satellite Factory Production Program, which fosters the establishment and development of subcontracting networks mainly among Taiwanese SME firms.

Type of Linkage Being Fostered

The backward linkages fostered in the four countries are all of a direct nature, mainly in the form of long-term, sustained local subcontracting. Their objective is not merely to increase the local added value and to have vendors produce parts and components based on well-established blueprints, but to foster an environment in which close working relationships between buyers and vendors are established so that true transfer and diffusion of technology and management know-how are possible. The foreign firms provide technical, managerial, and at times financial assistance to their domestic vendors and eventually cooperate with their vendors on the development of parts, components, and production processes.

In summary, the backward linkage policies of the four countries offer the following features:

1. They rely principally on market-oriented mechanisms.
2. In view of the nature of backward linkages, they are sustained and long-term oriented.
3. They enjoy a clear and shared understanding of their significance and objectives and have broad-base and high level political support.
4. They are closely linked to national economic development strategy and environment and consistent with related policies and programs.
5. They rely on a strong foundation of supporting programs, without which it would be difficult to enhance domestic firms' capabilities and to foster linkages.
6. They focus on targeted sectors that maximize linkage benefits or support industrial and import substitution policies.
7. Within targeted sectors, they make judicious use of limited resources and target a subpopulation of foreign and domestic firms, using the "winner strategy" approach.
8. They are stand alone or subsumed under the SME policy, depending on the strength of FDI presence and of the SME policy.
9. They seek close, long-term buyer-vendor relationships to allow sustained and effective transfer and diffusion of technology and management know-how.

PROGRAMS

In their implementation, backward linkage policies make use of an elaborate set of programs, divided into two distinct categories: those focused exclusively on backward linkages, and those that support SMEs.

Programs for Backward Linkages

Programs in this category are those established to deal directly and exclusively in making backward linkages happen. They exist in Singapore and Ireland—two countries with strong, stand-alone linkage programs, and in Taiwan where the program focuses also on linkages among domestic firms. All three programs, independent and separate from the SME programs, offer a number of functions, including

- Seeking specific backward linkage opportunities, usually within targeted sectors.
- Matchmaking individual firms for linkages.
- Advising domestic SMEs on business development and planning.
- Providing support to buyer firms to develop vendor network.

• Monitoring the progress of specific linkages and, if need be, trouble shooting those that are facing problems.

The three linkage programs are briefly described to give a flavor of what they do and how they function. They are: Singapore's Local Industry Upgrading Programme, Ireland's National Linkage Programme, and Taiwan's Center-Satellite Factory Production Program.

Singapore's local industry upgrading programme. The Local Industry Upgrading Programme (LIUP) ". . . is aimed at building up the efficiency, reliability, and international competitiveness of our supporting industry through forging close ties between local SMEs and MNCs operating in Singapore."[3] LIUP is an innovative support scheme initiated by the Economic Development Board to enhance the backward linkages opportunities with FDI firms and focus on supporting industry. The program is implemented in three phases: Phase I is to improve operational efficiency of participating SMEs; in Phase II, the product range of participating SMEs is widened and new production processes are introduced; and Phase III sees product or process development undertaken jointly by SMEs and their MNC buyers.

The LIUP consists of seconding for one year a manager from a participant MNC to the Economic Development Board and have him identify and select for intense and focused assistance through Singapore's SME programs a small number of local SMEs that have been or are potential suppliers to that particular MNC. Thus the LIUP offers participating MNCs the opportunity to develop their local vendors at the Economic Development Board's expenses. Participating SMEs are required to cover 30 percent of the assistance they receive under the LIUP, and the remainder is absorbed by government agencies.

The selection criteria for a SME to be part of the LIUP are stringent. A SME has to show a strong record of business performance and good business plans. Its management's attitude and receptiveness to the LIUP are assessed and must be found to be positive. It must exhibit capabilities for quality products and services, reliability, delivery, and cost competitiveness. In short, it must be attractive enough to appeal to the MNC as a partner eventually to conduct joint design and development projects in Phase III of the LIUP.

At the time of the survey, 22 MNCs agreed to cooperate with the LIUP and more than 65 SMEs are now receiving this focused assistance. By 1991 the Economic Development Board intends to increase the number of such SMEs to 100, or 10 percent of SMEs in supporting industry.

Ireland's national linkage programme. The National Linkage Programme (NLP) is formed by a consortium of five state organizations, with

Ireland's Industrial Development Agency in the leading role, to help local vendors ". . . become quality, reliable and price competitive suppliers to large buyers both at home and abroad."[4] The key operational objectives of the NLP are to identify industry linkage potentials within a selected group of sectors; to help develop a set of domestic suppliers; and to offer buyer support and development services.

The NLP has targeted five buyer sectors in which FDI has a strong presence—electronics, engineering, food and drinks, health care, and consumer products, and eight vendor sectors, in which Ireland has strong capabilities—print and packaging, sheet metal and finishing, automation, electronic manufacture assembly and system test, mold and die manufacturing, trade molding, light engineering, and food and fluid handling systems.

The NLP has identified 220 MNC affiliates for backward linkages in the five selected sectors; 70 percent of them have either committed to cooperate with the NLP or already had begun using domestic vendors. In targeting domestic firms, the NLP has adopted a "winner strategy." Of the 5,000-odd existing SMEs within the targeted sectors, the NLP works with 80 carefully selected SMEs to help them keep a customer (foreign affiliate) focus, resolve operational problems, access the full gamut of Ireland's generous SME assistance programs available to them, conduct business development activities, and finally enter into a local subcontracting arrangement with a foreign firm. The NLP provides services of a general nature to the remaining SMEs in the targeted sectors.

The NLP relies on no mandatory requirements to effect backward linkages and has combined all of the following functions and skills to attain its objectives:

1. Market research looks into supplying domestically the existing needs of foreign affiliates and seeks foreign buyers for existing domestic vendors.
2. Matchmaking works closely with vendors and buyers to remove all possible obstacles to reach subcontracting arrangements.
3. Monitoring and trouble shooting monitors the progress of on-going local subcontracting, and, with the express approval of the firms involved, acts as a trouble shooter when problems occur.
4. Business and organization development advises SMEs to help them build their management, production, accounting, quality control, and human resource systems, understand foreign firms' ways of conducting business, and develop business strategy and plans.
5. Broker for state assistance programs helps SMEs access the numerous and generous state assistance programs to enhance their technical, financial, and managerial capabilities.

It is noteworthy to mention the close and trustworthy working relationships the NLP has established with both domestic and foreign firms, whom it refers to as "clients." To be in a better position to serve foreign affiliates and understand their business and needs, the NLP succeeded in getting access to confidential information, such as their business plans and "spend lists." The NLP has established a reputation of honesty and of maintaining confidentiality, due to its professional business conduct, quality, and dedication of its staff and policy of providing excellent service to clients.

Established in 1986, the NLP is considered in Ireland to be an unequivocal success. For the 1986–89 period, over I£130 million worth of local subcontracting for foreign affiliates were achieved in the five targeted sectors as the direct results of the its activities. Its reputation is very high among both the foreign and domestic firms that have worked with it, and its services are highly valued and sought by them.

Taiwan's center-satellite factory promotion program. The purpose of the Center-Satellite Factory Promotion Program (CSFPP), established by the Ministry of Economic Affairs, is to organize and integrate factories (satellite) around a central one (center) to raise productivity and enjoy economies of scale, upgrade management and technology, create strong production and marketing networks, and enhance their international competitiveness. The CSFPP has developed three categories of networks, each with respective roles of the center and satellites:

Manufacturing:	Center	—assembly industries.
	Satellite	—manufacturers of spare parts and accessories.
Materials Processing:	Center	—producers of intermediate materials.
	Satellite	—processors of upstream raw materials.
Trading:	Center	—large-scale trading or turn-key export contractors.
	Satellite	—goods and equipment suppliers.

The Center-Satellite Force, set up to implement the CSFPP, supports center factories in establishing multilevel production systems; vendor assistance and productivity enhancement programs; and harmonious working relationships within the networks. Also, it assists the networks in working out a reasonable division of labor, based on factories skills and production capacities.

The Center-Satellite Force has tended to enlist larger and well-established factories from among SMEs to act as network centers. By October 1989, 60 networks were set up with 1,186 satellite factories in 15 different sectors, with a strong representation from the electronics industry.

All three programs focus on bringing about linkages between potential domestic SME vendors and foreign (domestic, in the case of Taiwan) firms buyers, with the ultimate objective of establishing long-term subcontracting arrangements. Although these linkage programs go about achieving their purpose in very different ways, they present common features worth noting. Each of these programs

1. Targets sectors and firms within those sectors as candidates for linkages.
2. Approaches linkages on a firm by firm basis to engineer a long-term, sustained buyer-vendor network.
3. Is founded on and makes use of a strong and comprehensive national SME program to enhance selected firms' capabilities.
4. Works closely with foreign firms (Singapore and Ireland) and displays professionalism and understanding of business.
5. Chooses a "winner strategy" approach.

SME Programs

The survey have shown that backward linkages efforts rely heavily on the SME programs, irrespective of whether the linkage program is stand alone or subsumed under it, or whether a hybrid or market-oriented approach to linkages is adopted. This is so because potential SME vendors have to be competitive and attractive enough on their own merits for foreign firms to consider them as vendors. Therefore a strong and comprehensive set of programs to assist SMEs develop their capabilities must be in place to remedy to their weaknesses and make linkage programs viable.

Though complex and varied, SME programs are designed basically to meet two sets of goals: to provide support and assistance to address the numerous technical, financial, managerial, human resource, and operational problems SMEs face in their daily business; and to promote and foster government policies and priorities by enticing SMEs with benefits to elicit desired behavior. The programs offer a host of financial subsidies and fiscal incentives, as well as technical advice and support, each meant to address a specific, often narrow SME problem or government policy or priority.

In view of their complexities and variety, and for our purpose, the discussion of the SME programs is limited to the type of assistance commonly found to enhance SMEs' business capabilities, without going into their tech-

nicalities. A sample of support for each category of assistance is listed below:

1. Technical assistance for productivity improvement, product and process development; and the transfer of needed or desired technology.
   Some programs include giving access to results of government-spon-
   sored research for their commercialization; coordinating similar technical efforts among different firms to obtain economies of scale and
   cross fertilization, and avoid duplication; and offering the services of
   government industrial laboratories to conduct tests and experiments.
2. Financial support in the form of access to financial markets and government credit guarantees, at favorable interest rates, for the expansion
   of production, the introduction of new products, and the management
   restructuring of the firm.
3. Management operations assistance to develop business, management,
   financial, and accounting systems; and to set up management information systems.
4. Business and commercial development to assist firms conduct market
   research and formulate business strategy and plan.
5. Human resource development to develop and train managerial and
   technical personnel.

Financial subsidies or tax incentives are attached to almost all of the assistance programs for SMEs. In most cases, a SME is required to cover a
portion of incurrred costs at a predetermined percentage or based on the
situation of the firm.

Many of the SME programs use consultancy services extensively, particularly in Singapore. The consultancy approach carries a number of advantages, not the least of which is the access to a wide variety of expertise at
favorable costs. Yet, one of its drawbacks is particularly felt among SMEs.
That is their inability to implement the consultant recommendations due to
the lack of either commitment by the firm, or internal technical expertise.

On the whole, the SME programs display a number of positive and practical characteristics. In most cases, the purpose and goals of each of the
programs are clear and precisely defined. Each is then designed and implemented in a professional and focused way, aiming at achieving, in some
cases with surgical precision, what it is supposed to. The programs tend to
be implemented impartially using transparent selection criteria. As a group,
they form a complementary, mutually supportive assistance system to address the gamut of issues related to the enhancement of SMEs' businesses.

Promotion of the Programs

Despite the coverage and generosity of many of the SME and linkage programs, it was found that SMEs are not taking advantage of them as much as they should or the authorities would like them to. A variety of reasons seem to explain this, including SMEs' lack of knowledge or true understanding of the programs, doubts about their benefits and effectiveness, and deep-seated apprehension at working with and opening up to the scrutiny of government agencies. On this last point, it was found that the younger and more educated manager/owners tended to be more amenable to tap into the SME programs.

To remedy these problems, promotional campaigns have been conducted to inform businesses about the programs and the benefits they offer. Singapore has adopted an innovative solution with the establishment of a number of Enterprise Promotion Centers in business and industry associations to locate the services of the programs out of government offices and closer to the firms, indeed in their own associations. There is evidence that the campaigns and other activities to promote the programs are paying off, as more firms become involved in both SME and linkage programs in Singapore.

INSTITUTIONS

In this section, the institutional setup for the backward linkage and SME programs is examined briefly to point to organizational factors and requirements that may determine their success. First, governmental and nongovernmental organizations responsible for the programs are presented, followed by a discussion of the coordination of their activities and their decision-making approach. Figure 5.2 provides a visual summary of the institutional structure.

Government Institutions

Linkage programs. The backward linkage programs in Singapore and Ireland, the two countries with stand-alone linkage programs and a strong FDI orientation, present totally different institutional setups. Singapore's LIUP is housed in the Economic Development Board and is staffed by a very small number of its professional employees. By contrast, Ireland's NLP, established by a consortium of five government agencies, is a stand-alone organization, though enjoying a very close relationship to the parent agencies, in particular the Industrial Development Authority. It has a professional staff of only 12, seven of whom are located outside Dublin, to be close to their

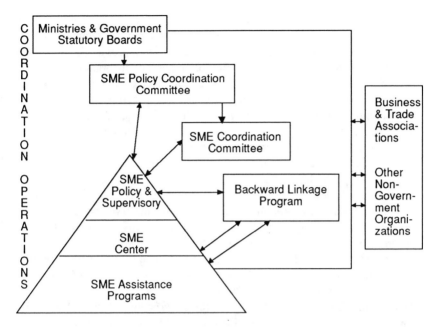

Figure 5.2. Backward linkage and SME programs institutional structure.

clients. To bring the staff together, jointly review projects, and share information and expertise, staff meetings are held on a regular basis. Apart from the NLP director, who was recruited from the private sector, the professional staff is seconded to the NLP by the parent agencies.

By its nature, a linkage program requires a staff that is technically competent, yet generalists and strategically minded to be in a position to deal with a variety of situations at the executive levels of domestic and foreign firms. As the result of their human resource policy, supported by an adequate budget, the two institutions seem to have attracted a dedicated, experienced, high caliber professional staff, with a good understanding of the issues, industrial sectors, business conduct, and government assistance system.

SME programs. Conceptually, the structure of the SME program is a three-tiered pyramid: as its base are the numerous assistance programs that SMEs tap into to develop and strengthen their capabilities; one level up is an operational body with the overall responsibility to implement the SME program; and on top is a policy- and strategy-oriented body that wields overall management and control of the SME policy and programs. A brief introduction of each of the levels follows.

At the lower level, the SME assistance programs conduct a wide variety of support activities, located in the government agencies technically best qualified to manage them. For example, programs that provide financial and fiscal benefits are located in ministries of finance or in organizations empowered by them. Those dealing with industrial and technical support belong to ministries of industry, productivity centers, and institutes of economics and technology. Those providing more technically oriented assistance, including R&D, technical standards, and similar types of support are located in ministries of science and technology, standards centers, and research centers. A variety of training centers offer human resource development and training at all personnel levels.

At the middle level, each of the four countries has established a SME center that has overall operational responsibility for the SME program. SMEs go to the centers for inquiry and advice on the programs. The centers devise a program of assistance for a particular SME and coordinate its implementation to ensure that duplication is minimized and that support needs are met. Some centers even provide SMEs with a service as to how to fill out applications and prepare supporting documents.

Here again, the institutional setup for the SME centers varies substantially from one country to the other. In Taiwan, the Medium and Small Business Administration, located in the Ministry of Economic Affairs, acts as the SME center. Similarly, South Korea's Small and Medium Industry Promotion Corporation is housed in the Ministry of Trade and Industry, with the difference that it is legally a government-owned corporation. Ireland and Singapore's SME program are located, respectively, in the Industrial Development Authority and the Economic Development Board, two powerful, semiautonomous government organizations that have aggressively promoted economic development. As mentioned earlier, Singapore has spun off much of the promotion and first contact functions of the SME center into Enterprise Promotion Centers located in business and industrial associations.

The SME policy- and strategy-making body responds to high level government requests to develop policy initiatives, designs and plans programs, coordinates the activities of high level government agencies involved in the SME policy and programs, supervises and controls the operations of the SME centers, and get involved in related major strategic decisions. In all cases, these bodies are bureaus and divisions in the government organizations to which the SME centers belong. For example, the Small and Medium Industry Bureau carries these functions within South Korea's Ministry of Trade and Industry, and supervises the Small and Medium Industry Promotion Corporation, the country's SME center.

Nongovernment institutions. The backward linkage and SME programs in all four countries consider important and desirable the private sector's involvement in the programs, because their mission is to develop the private sector and make it more competitive. So it is a sign of strength that the private sector has forged a partnership with government agencies to make a positive contribution to the programs and that SME associations are politically powerful, resource rich, competent, and nationally representative. The case of the Korea Federation of Small Business (KFSB) is illustrative.

The KFSB represents over 300 SME cooperatives, encompassing over 23,000 SMEs, receives substantial government subsidies amounting to half of its annual budget, and is considered one of the four most powerful business associations in South Korea. The Federation's role in the SME program is substantial and influential. It represents its members' interests to government, both on the policy and operations levels, through regular meetings with agencies involved in the SME program and monthly meetings with the president. It has been empowered to manage certain SME programs and to review local subcontracting agreements to ensure that SMEs' interests are preserved according to law. The Federation has in-house advisory services to help SMEs modernize and access SME assistance programs and to promote SMEs' involvement in foreign trade and investment. Its many publications, including a weekly newspaper, form a major source of information about the SME program.

Other nongovernment organizations involved in the linkage and SME programs have included business associations, research institutes, universities, consultants, and consulting firms. Their participation has been mainly in the operations of programs.

Institutional coordination and structure of decision making. Both the linkage and SME programs directly involve a large number of organizations across a variety of disciplines. Their success depends on the cooperation and contribution of those organizations in the design, formulation, and implementation of the linkage and SME policies. It is clear that no single agency is in a position to devise the programs and implement them without other agencies' cooperation and support. It is also clear that a multiagency network has to be put into place to genuinely elicit and coordinate those organizations' participation and contribution.

Institutionally, in Singapore two committees fulfill that coordinating function: at the policy level, the Committee on Small Enterprise Policy is chaired by the minister for Trade and Industry and comprises the permanent secretaries of selected ministries and the heads of statutory boards; on the working level, the SME Committee is led by the Economic Development Board and includes representatives from other government agencies. As mentioned, Ireland's NLP itself is formed by a consortium of five government agencies, each one of which second one or more of its staff to the program.

The decision-making structure of the programs must reflect diverse factors. On the one hand, there is the need for organization leadership and strategic and policy-decision making authority. On the other, a large number of government units located in a number of agencies are involved in the daily operations of the programs. Each has its own expertise and must be responsive to their clients. To be able to address their needs, all linkage and SME programs have displayed an interplay of centralization and decentralization of decision making.

For example, Ireland's Industrial Development Authority, Singapore's Economic Development Board, and South Korea's Small and Medium Industry Bureau centralize decision making at the policy and supervisory levels. Most of the operational decision-making structure is decentralized. For example, in Singapore, much of the program promotion activities and client relations and services are being spun off to business and trade associations. Assistance programs provide even a more striking example of decentralization. A number of programs have established "approved-in-principle" mechanisms, based on clearly stated conditions and criteria. This reduces unnecessary intervention of central government units in routine, simple, and operational decisions and allows a faster response rate to SMEs.

CONCLUSION

The experience of countries that have established policy and programs to foster backward linkages has shown that two philosophically different approaches have been adopted. The first, the administrative-oriented approach, relies more on legislated import substitution and high trade barriers to effect the linkages. When not coupled with mechanisms to spur efficiency, for example, through competition, this approach has not been very successful.

The other, market-oriented approach, holds the view that linkages should be achieved mainly on an economic and business basis, with the free will of foreign and domestic firms to enter into them. This approach, by far the more complex, is relatively new and exists in a small number of countries to the extent described here. It seems to hold much promise, as the survey has shown.

The concept of backward linkages is simple, yet the design of an effective policy is complicated and a rather difficult exercise. This is because a backward linkage policy must fit in with the national economic environment and development strategy, as well as be compatible with related policies. This implies that the design of the policy is bound to create situations of conflicting interests, where higher-ups will have to arbitrate and make decisions and compromises.

On the programmatic side, the study has clearly shown the importance of establishing a backward linkage program that works closely with both foreign and domestic firms, on a firm by firm basis to achieve one linkage after the other. The linkage program's key role is that of a matchmaker, trouble shooter, and provider of access to assistance programs.

The linkage program is not viable without the presence of a well-established, comprehensive, and effective system of assistance programs to SMEs, which seem to be the linkage program's almost exclusive point of focus among domestic firms. This system of assistance programs, used to enhance SMEs' capabilities to render them more attractive for local subcontracting, is a precondition to the establishment of a backward linkage policy and program.

The institutional structure of the backward linkage and SME programs reflect the complexity of the linkage and SME policies. It is thus essential that the necessary political support be present and the organizational mechanisms be put into place to establish a multiagency network and obtain an effective coordination among its members, both crucial to the success of the linkage program.

Using a programmatic and institutional approach, this chapter analyses the experiences of Taiwan, South Korea, Singapore, and Ireland in fostering backward linkages of FDI into the domestic economy, as a major mechanism to the transfer and diffusion of technology. Clearly, there are other major dimensions of direct impact on effecting backward linkages not covered here. For example, because the economies of all four countries enjoy a certain developmental level and their bureaucracies are known to be efficient and generally noncorrupt, to what degree are the findings relevant to countries with a lower economic development stage and a less efficient bureaucracy? The backward linkage programs surveyed to focus on a rather narrow range of manufacturing industries, with discreet production processes. How would they be affected if industrial sectors vary more widely? How do factors ranging from shifts in the global business environment to particular MNCs' production and supply strategies affect backward linkages? In the design and implementation of market-oriented SME and linkage programs, how to minimize market distortions so as not to support economically nonviable SMEs and linkages? All these and other dimensions deserve to be examined to promote our understanding of backward linkages as mechanisms for the transfer and diffusion of technology.

ACKNOWLEDGMENTS

This chapter is based on a four-country survey sponsored by the Foreign Investment Advisory Service (FIAS), IFC/MIGA, the World Bank Group.

The author would like to thank FIAS for granting permission to reprint a version of the main findings of the survey published by FIAS. The views expressed here are the author's and do not represent those of FIAS.

Notes to Chapter 5

1. Backward linkages as mechanisms to the transfer of technology have been illustrated in many studies, including: Axel J. Halbach, *Multinational Enterprises and Subcontracting in the Third World: A Study of Inter-Industrial Linkages*, working paper no. 58, ILO, Geneva, 1989; *Transnational Corporation Linkages in Developing Countries: The Case of Backward Linkages Via Subcontracting—A Technical Report*, Center on Transnational Corporations, UN, 1981; *Subcontracting for Modernizing Economies*, UNIDO, 1974; Hal Hill, "Subcontracting, Technological Diffusion and the Development of Small Enterprise in Philippine Manufacturing," *Journal of Developing Areas*, January 1985.

2. For the purpose of this chapter, unless otherwise indicated, backward linkage is intended to mean linkages between foreign affiliates and domestic firms.

3. *SME Master Plan—Report on Enterprise Development*, SME Committee, Economic Development Board, Singapore, May 1989, p. 62.

4. Presentation made by the NLP, Dublin, October 1989.

Chapter 6

Industry Research Associations: The Canadian Experience

Isaiah A. Litvak

Canada's resources for scientific and technological research are thinly spread and limited. In fact a significant feature of the Canadian corporate landscape is that the great majority of its companies do no research and development (R&D) at all. In 1987, 100 companies accounted for 74% of the R&D performed in the "Business Enterprise" sector, and the remaining 26% was performed by some 1,700 other firms.[1] If Canadian companies are to be competitive at home and abroad, they must access both relevant and emerging technologies. Most Canadian companies, however, do not possess the resources to develop and/or acquire state-of-the-art technologies through their own resources and efforts.

In 1988 the Science Council of Canada recommended that greater consultation be encouraged among government, universities, and business in targeted industries. A key objective of the consultation process was the promotion of greater self-help by industry in the areas of R&D and technology.[2]

Whereas the concern with promoting research and development and technological innovation has been on-going, it is only in recent years that Canadian governments have stressed the need to encourage trade associations and industry research associations (IRAs) to perform a more active role in the promotion and diffusion of technology.

The purpose of this chapter is to highlight the reasons that the Canadian government is committed to encouraging the development of IRAs as an instrument for promoting collective R&D initiatives; accessing domestic and foreign technology sources, particularly on behalf of small and medium-size enterprises (SMEs); and promoting the establishment of industry alliances involving business, government, and university scientific personnel and laboratories. As of 1990 Canada has approximately 10 active IRAs. Industry rationale and government support for establishing two of them are presented here in the form of vignettes. They are among the more active IRAs in Canada.

CANADA'S TECHNOLOGICAL POSITION

Based on international comparisons of research intensity with other industrialized countries, Canada fared badly in the 1980s. Its ratio of gross expenditures on R&D to gross national product was 1.4 percent in 1986, the lowest among the Group of 7 leading industrial nations (G-7). The performance of the other six countries was, in rank order: Japan—2.9% (1987), West Germany—2.8% (1987), United States—2.6% (1987), United Kingdom—2.4% (1986), France—2.3% (1987), and Italy—1.5% (1987). Equally important, Canada's total R&D expenditure at U.S. $5.6 billion in 1986 pales in comparison to that of the United States at U.S. $117.4 billion, or that of Japan at U.S. $41 billion.[3] In fact, Canada did badly in all important measures of technology performance. Its poor technology record is in no small measure responsible for its dismal performance as an exporter of high technology products, that is, lower than for any other industrialized country.[4] However, concern with Canada's technological performance is not limited to the high-tech sector.

A major weakness in Canada's technological policy is the focus on domestic production of new technology while the processes of adoption and diffusion are neglected. In various studies, the Science Council of Canada and the Economic Council of Canada concluded that investing in scientific R&D was not enough and that attention should be given to additional activities such as improved mechanisms for the transfer of technology.[5] In 1980 the National Research Council warned that:

> Canada's total output of technology amounts to less than one percent (1%) of the total world output . . . It is a matter of considerable urgency that efforts be made to bring the ninety-nine percent (99%) of world technology forcefully and more conveniently to the attention of the possible exploiters, which are mainly to be found in industry.[6]

Some nine years later, the Natural Sciences and Engineering Council warned that if Canada were to compete in the global economy:

> . . . it must develop and maintain the ability both to produce a broad range of research driven technologies and also to identify, obtain and adopt tech-

nologies produced elsewhere. These technologies are needed to strengthen and maintain our competitiveness in our mainstay resource industries (for example, forest and food products), to develop the new technology intensive industries (for example, telecommunications), and to establish a more ecologically benign industrial base.[7]

TRADE ASSOCIATIONS AND THE DIFFUSION PROCESS

Interest has been expressed in the role of trade associations and IRAs in enhancing the diffusion of technological and managerial advances in Canada, but the effectiveness of association in doing so has not been substantiated. Recommendations made by the Economic Council of Canada centered on the need for associations to become responsible for collecting and disseminating "information on new ideas and best-practice technology and management methods in use in Canada and abroad"[8] and that government financial assistance be provided to help.

The precedent for this type of activity and government support is to be found in a number of European countries. In France, for example, "cooperative industry research institutes have aided in the development and dissemination of technology, and have a mandate to provide training, advice, and information to member firms."[9] The social benefits derived from the dissemination of such information to private firms are deserving of some public financial support, according to the Economic Council of Canada.[10]

The Science Council of Canada also made recommendations regarding associations—that associations should establish frequent contact with government laboratories at all levels and should press their members to do the same.[11] Associations are seen as a necessary element in technology transfer because outward diffusion of technology to the membership is not the only requisite, but collection and interpretation of members' needs to present a unified voice to the federal laboratories is also important.[12]

A subsequent study published by the Science Council of Canada recommended that sectorally oriented technical centers, established and supported collectively by the industries that would use them, are needed in Canada. The authors reasoned that the capacity for technology absorption at the enterprise level should be increased. Pointing to the example of European sector-oriented technical centers, they argued that such centers would be knowledgeable about problems faced by small firms in their respective industries, and so could act as technological interfaces between those firms and the many sources of technology and information.[13]

Successive Canadian governments have reiterated the technological challenge facing Canadians; namely, the need to improve Canada's capacity to develop/apply new technologies in an increasingly tough competitive global marketplace. Policymakers acknowledge that the Canadian reality makes it more expedient and desirable to adopt technology than to create it anew. To this end, many recommend that the research work of government laboratories be made more sensitive to the needs of industry and that trade associations along with other organizations be part of an active consultative and transfer mechanism for making government laboratory research more relevant for Canadian business.

INDUSTRY RESEARCH ASSOCIATIONS PROGRAM

The Industrial Research Associations Program (IRAP) was the first government initiative aimed at helping to finance the establishment and promotion of nonprofit organizations that carry out and/or facilitate R&D work related to the problems of an industrial sector.

In the 1970s and early 1980s, support for IRAs amounted generally to $200,000 per annum for periods of 5 to 7 years. Although not all IRAs applied for IRAP assistance, all benefited from some form of government assistance and support. Nonetheless, the expectation was that the IRAs would become financially independent of government funding and be able to sustain future operations through contributions from supporting companies and fees for service provided under contract.

The rationale underlying government support can be gleaned from the following excerpts taken from a letter written by a government official:

> IRAs have a number of characteristics that make them an attractive mechanism for technology transfer and industrial development. They foster a greater cooperation between members of a particular industrial sector and the creation of cooperative research programs; enhance the links between technology and R&D institutions, governments, universities and industry; facilitate the transfer of technology to and the diffusion of technology across an entire sector of the economy; provide valuable training services to their members; and, generally increase the technological knowledge and awareness through activities such as contract research and seminars . . .
>
> From [the government's] perspective, IRAs are one of the best instruments for establishing an industrially relevant technology centre. The industry commitment (technical and financial) to the IRA, along with the business focus that is imposed in the management by the industrial members, ensures that the institution will perform services that are of value and importance to the industrial community and that it will strive towards self-sufficiency at an early stage.[14]

The government has since replaced IRAP with TOP—Technology Outreach Program. TOP's primary goal is to help improve the productivity and competitiveness of Canadian industry by providing a supporting infrastructure to accelerate the acquisition, development, and diffusion of technology and critical skills in Canadian industry, with a particular focus on SMEs. Industry Research Associations are eligible for TOP support.

Certain IRA startups that received support under IRAP have also received substantial funding under TOP. The Canadian Plastics Institute, for one, in 1988 received $1,900,000 over five years to assist it "to diffuse leading-edge plastics processes and product technology to the plastics industry, improve access to information on product design, materials and process selection, quality control, testing and inspections." The two vignettes that follow illustrate the nature of government IRA support.

THE CANADIAN PLASTICS INSTITUTE

The movement to establish a Canadian Plastics Institute received its formal start in the mid-1970s. Studies conducted by government and the Society of the Plastics Industry of Canada (SPI)—the national trade association whose members include plastic resin manufacturers through fabricators of plastic products and including makers of machinery, molds, and auxiliary materials used in the manufacture of plastic products—indicated that the Canadian plastics industry was behind the American, European, and Japanese industries regarding early knowledge of technical developments.

Small firms were found to be at a greater disadvantage than larger companies because the latter could make use of technology transfer from affiliated or associated companies. In addition, the problem was compounded for the average Canadian plastics processor who could not justify the expense of research and development activity because of the relatively small size of the Canadian market.

The government viewed plastics as a priority industry.

> Plastics are fundamental to the international competitiveness of most manufacturing sectors today. They are the most prominent of the advanced industrial materials and they will become of greater strategic importance as new applications are developed. As most nations have growing and developing plastics industries, it is of primary importance to ensure that the plastics products industry in Canada is internationally competitive.[15]

The strategic importance of the plastics industry may be attributed in part to the role of polymer technology as an enabling technology.

In May 1979 the federal government agreed with an industry sector task force that recommended financial support be given to establish a research institute. The SPI model for such an institute was taken from a Science Council report that described the function of an institute as improving the productivity or operational efficiency of industry firms by

- Acting as an interface between these firms and sources of new technology.
- Adapting new technologies to the needs of these firms.
- Assisting firms to improve upon their use of existing technology and management.
- Undertaking development activity and some research relevant to the industry.
- Offering standardization and testing services.[16]

The Canadian Plastics Institute (CPI) was established in Toronto by SPI on March 1, 1983. In total, eight years and four reports elapsed between SPI's initial recommendation to the government and CPI's actual start. Along the way SPI received considerable federal and provincial government support. Of particular importance was the federal grant of $200,000 per year for five years, which was promised to CPI for its startup under IRAP.

The major barrier to CPI's establishment was internal to SPI—there was resistance to the idea from the membership. Large firms that had their own R&D capabilities could not see the benefit of having another such organization. Most of these firms were foreign-controlled. Fear of competition kept other SPI members from accepting the idea because they did not want to see competitors improving their technical and managerial capabilities. Although the SPI board supported in principle the idea for CPI from the start, it took much selling by SPI's president to convince the general membership of its worth.

During that eight-year period, the board's concept of CPI's function remained the same: to improve productivity or operational efficiency of customer firms. Some activities initially proposed for CPI, such as in-house R&D and testing services, presumably to lessen the risk by decreasing the amount of investment required, were reduced with time.

One major theme seemed to underlie the drive to establish CPI over the entire period—Canadian plastics products were in an unnecessarily poor position to compete internationally in terms of both cost and quality. Technology was seen as the key to achieving cost reductions and quality improvements. Because the majority of plastics firms were too small to conduct their own R&D and because an enormous amount of undigested information was in the public domain, the emphasis was on improving the diffusion of technology throughout the industry. The Canadian Plastics Institute was seen as the ideal means to do it.

The two mandates given to CPI at the time it was established were: "to act as an information processing service and to develop and utilize a network of laboratories and other resources" and to provide "technology support for improving existing products and developing new plastic products and more efficient production techniques."[17]

To carry out this work, data bases from around the world were to be accessed; a technical information officer was to be hired to develop CPI's own data base and manage all other information; and relationships were to be established with the laboratories of the provincial and federal government research organizations, universities and colleges, private commercial research organizations, and, where appropriate, corporate laboratories. CPI has no laboratory facilities of its own.

The first CPI president was appointed in November 1983. Some seven years later, in 1990, CPI's mission is still "to promote technological advance in support of the international competitiveness of the Canadian plastics industry." CPI has a staff of six plastics industry experts, two of whom are industrial technology advisors funded by the National Research Council of Canada. CPI's activities can be grouped as follows:

1. Technology acquisition. The objective is to locate technology that is ahead of current North American practice; ready, or almost ready, for commercial use; and available for licensing or joint-venture arrangements to Canadian firms for commercial exploitation. A key aim is to facilitate the establishment of strategic alliances between Canadian manufacturers and processors and compatible foreign and domestic entities.

2. Technology dissemination. Through seminars, publications, and direct liaison with companies, CPI provides Canadian companies with information regarding global developments in plastics technologies. CPI staff regularly visit SMEs to provide technical information and to assist them in identifying market opportunities and accessing new sources of technology.

3. Technology implementation. A key goal is to maintain contact with companies trying to adopt technology and offer technical assistance in choosing the best materials and processes, and in improving new products. Smaller processors are more responsive to employing CPI's services because they lack resources to keep up with global developments and trends.

4. Facilitating and fostering research. CPI is committed to promoting dialogue and working relations between the global research community and the Canadian plastics industry by participating in Canadian and

international research organizations. It also strives to encourage university scientists and industry researchers to exchange information and match their research goals and agenda more closely.

To ensure that CPI continues to carry out its role as an agent for technological development, in 1988 it received a five-year government commitment worth $1.9 million under TOP. Other federal and provincial government support has been received in the form of grants for research and travel, in addition to professional and technical manpower assistance. It is unlikely that CPI would have survived without government assistance. Equally important, current fees from publications and client services could hardly support the CPI staff complement and related expenses incurred in the pursuit of its mandate.

THE CANADIAN STEEL INDUSTRY RESEARCH ASSOCIATION

The Canadian steel industry is regarded as one of the most efficient and profitable in the world, in spite of being limited to a small domestic market and its scale implications. The industry is Canadian-controlled, and government ownership and direction are not significant compared to the role played by governments in other producing countries.

R&D expenditures as a percentage of sales for the steel industry is lower in Canada than either in the United States or Japan. There is virtually no fundamental research on a laboratory scale conducted within the Canadian steel industry. This is in sharp contrast to the situation in Japan where certain steel companies have corporate laboratories dedicated to conducting studies of a fundamental nature.

What has helped the Canadian steel industry to succeed is its commitment to acquire the most up-to-date technologies at the earliest point possible. The two key technological developments in steelmaking—the basic oxygen furnace and continuous casting—were both pioneered in North America by Canadian firms, Dofasco and Atlas, respectively. As the Canadian steel industry is small by world standards and innovation is necessary to remain competitive, the industry recognized the need to cooperate in certain areas of steel research.

The Canadian Steel Industry Research Association (CSIRA) came into being in June 1978 to promote iron and steel research through improved communications and pooling of effort, as follows:[18]

1. Canadian governments
   (a) Provide an industry viewpoint on the orientation of iron and steel

research programs conducted by government laboratories and agencies.

(b) Assist and encourage government laboratory and agency heads to transfer technical information to association members.

(c) Provide input regarding new or proposed relevant legislation.

(d) Assist member companies in dealing with the political and administrative branches of government on research-related matters.

2. Association member companies

(a) Identify the needs for technological improvement within the Canadian steel industry.

(b) Provide a forum for technical interchange.

(c) Promote the formation of research task groups.

3. Institutions

(a) Guide and encourage research in universities and colleges toward the needs of the Canadian steel industry.

(b) Encourage the members of the association, either independently or collectively, to sponsor research at universities and colleges.

(c) Influence the curriculum of universities, colleges, and technical schools to meet the needs of the Canadian steel industry for qualified personnel.

4. Other associations

(a) Provide liaison with other national and international associations.

In essence, the primary motives for establishing CSIRA were

1. To redirect government laboratory and university research toward the needs of the iron and steel industry, and

2. To influence government research incentive programs so that industry could effectively utilize them

CSIRA membership is limited to Canadian-based steel firms having both melting and rolling facilities. Each member firm may appoint one person, usually from their R&D or quality control departments, to represent it on CSIRA's board of directors. From among its members, the board elects a chairman, vice chairman, and treasurer. CSIRA's only employee is a part-time technical director, whose responsibilities include setting up committees to work on various aspects of subjects of common interest, shepherding technical studies approved by the board, and preparing minutes of all board meetings.

CSIRA's 1988–89 administrative budget was less than $100,000. All revenue came from membership fees and investment income, about 80 percent of which went toward the technical director's salary, publications, and overhead. Specific activities such as research projects, studies, and workshops receive funds from government sources as well as additional member contributions.

Although CSIRA's administrative budget is small, it has a good deal of influence on steel research in Canada. The membership identifies R&D problem areas common to the industry in terms that are suitable for research. The scale of these problems is generally too large to be undertaken by an individual firm, and they are directed to university or government laboratories to carry out. Steel R&D takes place within university, government, and industry facilities. In the case of university research, CSIRA may support the university's application for government funding. In some cases CSIRA will provide the seed money for studies to be carried out using members' facilities and/or consultants.

To strengthen the ties between the Canadian steel industry and government research laboratories, CSIRA and the federal government's Canada Center for Mineral and Energy Technology (CANMET), signed a memorandum if understanding creating a formal framework for collaboration and cooperation. One of CSIRA's objectives is to maximize economic benefit from government funding of research. To do this, increasing emphasis is placed on communication with government. Periodically CSIRA identifies what it regards as industry's technological needs in both process and product technology areas. This information is made known to publicly funded agencies.

Inasmuch as technology transfer is a two-way process, CSIRA places great importance on promoting "effective communication" with outside agencies. Conceptually, CSIRA's communication strategy can be viewed as shown in Figure 6.1.

One measurement of performance employed by CSIRA is to determine the extent to which its communications strategy results in improved "technology transfer, that is, university/government-developed technology to the steel industry."

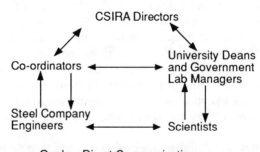

Figure 6.1. CISRA's communication strategy.

Another CSIRA objective is to provide seed money and/or facilitate the award of research grants to key individual investigators, largely at universities. Some research leverage is occasionally realized by becoming supporting members of larger research projects. The greatest part of these efforts lies in applied research and development. Little basic research is funded, although CSIRA does support some studies of fundamental interest. In contrast, little product development is funded, because such research activity is usually left to the competitive initiative of individual member companies.

It is not necessary for CSIRA's 13 members to be involved in all association research initiatives and programs. The organization possesses the flexibility to accommodate the needs and capacities of its different members. The six largest members, for example, formed a research consortium and incorporated a separate company—Project Bessemer, Inc.—which has as its mandate to develop strip casting technology. The sheet steel research is expected to cost more than $10 million. Experts from Canadian universities will provide scientific assistance as many aspects of the technology are unknown and require intense fundamental research. Concurrently, scientists in government laboratories are engaged in applied research for immediate use in the project. Although the Bessemer project was spun off from CSIRA as a separate entity, the association retains its interest in near-net-shape casting as a continuing facet of its strategic research plan.

FINDINGS

In addition to researching the activities of CPI and CSIRA, the findings that follow also draw on information obtained from two other active IRAs— Canadian Gas Research Institute (CGRI) and Pulp and Paper Research Institute of Canada (PAPRICAN).

National Trade Association Links

Canadian national trade associations gave birth to three of the four IRAs. The Canadian Pulp and Paper Association established PAPRICAN; CGRI's roots belong to the Canadian Gas Association; and the Society of the Plastics Industry of Canada was instrumental in setting up CPI. The IRA boards were dominated by association members. Membership dues and contract research constituted the bulk of nongovernmental financing, and industry members played a critical role in formulating IRA research agendas.

At the time the IRAs were formed, all three national trade associations had a technical division/section in their organization. The responsibilities of these technical units included: product testing, maintaining links with government laboratories and universities, sponsoring and conducting scientific and technical symposiums, helping to subcontract research projects on behalf of industry members, etc. In fact, these units performed the technology "gatekeeper and brokerage" role for association members. Thus it was not surprising that national trade associations that had technical units were the most likely candidates for spinning off IRAs.

CSIRA was the only IRA that was not a trade association spin-off. In fact, it came into being before a Canadian Steel Association was established. The seven major Canadian steel producers were members of the American Iron and Steel Institute, a large American association that also included a "technical" unit. The decision to form a Canadian IRA can be attributed to the steel industry's desire and need: to promote research into steel at Canadian government laboratories, universities, and other scientific institutes; to encourage educational institutions to graduate professionally competent personnel for the industry; and to influence public policy in R&D in favor of the Canadian steel industry. The foregoing concerns were equally shared by all Canadian IRAs.

Government Support and Involvement

The federal government played a key role during the start-up phase of each of the four IRAs. In fact, government financial support and encouragement was instrumental in the setting up of IRAs. CPI was established in 1983 with the help of a federal grant of $200,000 per year for five years; CGRI's start-up in 1974 was also assisted by an IRAP grant of $175,000 per year for the first five years of its operation; PAPRICAN, Canada's oldest and leading IRA, when formed in 1925 was given $36,000 a year, for a four-year period; and although CSIRA did not get a grant per se, it received professional manpower assistance at the time of its formation.

Other types of support included: partial financing of consulting studies designed to assist the trade associations to determine the most suitable IRA organizational form as in the case of the CPI; giving capital grants to PAPRICAN in the form of operating facilities, including laboratory equipment; placing contract research work with both CGRI and PAPRICAN; and helping underwrite projects undertaken by IRAs such as CSIRA's product quality sensor project. Government officials were involved in some working capacity in all four IRAs, ranging from ex-officio board appointment to serving as secretary.

Relations With Governments and Universities

All four IRAs had working relations with federal government departments and laboratories, with universities, and to a lesser extent with provincial governments and their research organizations. Equally important was the fact that there was interchange between IRAs and their scientific and technical counterparts in the government sector and university community. PAPRICAN's network was the most extensive and included a cooperative program of graduate education whereby select graduate students at McGill University and the University of British Columbia carried out their thesis work under the direction of institute staff members.

The president of CPI hoped to rapidly expand links with government, university, and other public-funded laboratories by being appointed to the advisory board of the National Research Council's Industrial Materials Research Institute (IMRI). These links formed a resource network to be tapped by association members. Comparable network strategies were employed by CSIRA, CGRI, and PAPRICAN.

Similarly, the IRAs had ex-officio members on their boards/committees drawn from government departments, labs, and agencies. For example, the director of the Metals Technology Laboratories at CANMET and a representative of the Resource Processing Industries Branch, Department of Regional Industrial Expansion were members of CSIRA's board.

Of the four IRAs, PAPRICAN played the most critical role in supplementing member research programs. Corporate member participation in determining PAPRICAN's research agenda and the size of the budget assigned to in-house research were two important reasons why this IRA had contributed to the growth of the Canadian pulp and paper industry. Unlike other Canadian IRAs, PAPRICAN has had long, well-established, and close relations with university and government scientific communities in Canada and abroad. The combination of in-house research and the wider diffusion of research undertaken by outside organizations and individuals had allowed PAPRICAN to play a critical role in the transfer of technology on behalf of its members and industry.

By collaborating with universities in graduating masters and doctoral students with research interests in pulp and paper, PAPRICAN has helped provide a steady stream of high quality scientific and technical personnel who have benefited from interaction with industry researchers. Some of these graduates have conducted research considered significant by industry R&D leaders. In recent years, PAPRICAN has expanded its educational role to include training technical personnel for industry members.

Industry Base

The four IRAs operated in industries that were considered vital to Canada's economic health—plastics, steel, gas, pulp, and paper. Although not of the high technology type, continued productivity improvements had to be realized if Canadian firms were to survive in an increasingly competitive global marketplace. The gas and pulp and paper industries were among Canada's leading exporters; to be profitable, the steel industry had to be internationally competitive; and the plastics industry had to improve its technological competencies to meet international competition.

As noted, none of the industries was technology intensive with less than 0.6 percent of industry sales going to R&D. For this and related reasons, a critical component of the collective research undertaken/subcontracted by the IRAs was process-oriented with the goal of developing technologies that would lower costs and increase productivity vis-a-vis international competition. At PAPRICAN, for example, basic in-house studies on calendering led to the development of techniques that were applied in the industry, assisting member companies to make design and operations decisions with respect to their commercial calendering stocks.

For an IRA to be successful, it must be industry-oriented and supported. To have the support of industry, it should be governed by member company representatives. A key challenge facing IRAs is how best to serve the needs of their corporate members, which are often diverse and in large measure reflect differences in company size and technological capabilities.

Large companies, for example, tend to have an in-house technological capability and their outside requirements are usually of a more sophisticated nature. This observation applied equally to Canadian and foreign-controlled firms. Small member companies without R&D facilities and personnel, in contrast, have needs that typically involve the acquisition of existing technical information, product testing, or other routine technical activities. The member group in the middle includes companies that, although not having their own R&D facilities, employ some trained technical personnel and have the financial resources to engage extramural R&D assistance to generate and supply them with proprietary technology.

Firms that conduct R&D tend to have greater interaction and closer working relations with IRA staff. This fact favors the larger firm, whose R&D personnel are in a stronger position to condition the IRA's research agenda.

> . . . many research institutes appear to have been 'captured' by the larger, more technically capable firms, who really have less need for their services. There is a very high cultural barrier deterring interaction between institute staff and those smaller firms with less technical capability, whom they are really supposed to serve. . . . The smaller firms with few, if any, technically trained personnel have great difficulty understanding the work of a research institute, and have an understandable reluctance to even attempt to communicate with such an institution. There is a perceived barrier that prevents firms, particularly small backward firms, from using this potential source of technology.[19]

SOME FURTHER OBSERVATIONS

The operating presence of only a handful of private sector-based IRAs is an indication that significant constraints exist in Canada to this type of organization as a vehicle for promoting technological innovation. In fact, some IRAs are operating on government life-support systems, most do not have a strong enough revenue base to plan medium-term, let alone long-term strategic research programs, and all are constantly having to lobby their membership and government for additional financial support and assistance.

Canadian industry does not appear to be motivated to promote collaborative ventures in technology. Initiatives, when they do emerge, result largely from the availability of government assistance coupled with strong leadership and support by association executives. The culture of the private sector militates against a collectivist approach. Large firms prefer to do their own R&D, and small firms are generally too suspicious, ill-informed, and individualtistic to agree to enter into collaborative arrangements. Hence, it is not surprising that much of IRA research is process-oriented.

Comments made by trade associations and IRA representatives concerning the state of technology and industry competitiveness in Canada suggest possible ways in which industry associations could play a significant role. A number of their observations have one thing in common—lack of awareness among producers or users of technology. There is evidence that many manufacturing firms are ill-informed about the roles of government laboratories, that transfer of technology to industry is not perceived by most of the federal labs to be one of their primary missions, that inter- and intraindustrial linkages involving technological developments are weak, and that laboratory awareness of the work going on in other labs is poor.

Because industry associations represent existing mechanisms that link firms in the same industry, they appear to be appropriate means for increasing awareness of technological advances within an industry. The association's communications network could be used to educate members as to what government and university labs and technology centers could do for them and to encourage members to pressure those labs and centers to increase their efforts at technology transfer. The industry association should be in the best position to understand the research needs across its industry and so could act as the link between labs and centers with respect to research relevant to that industry. The interindustry link could be forged through relations between associations.

A major problem experienced by small business in Canada was one of insufficient or lack of technical expertise. This problem can better be defined as one of "corporate resource poverty" in such areas as research and development, industrial engineering, and state-of-the-art management practices.

IRAs in particular can play a vital role as knowledge networks for small and medium-size firms by facilitating technology transfer through the provision of available technical information, advice, and know-how. In addition, the IRAs may actually carry out research and/or development on a "a client for a fee" basis involving multiclient groupings via collective industry research projects of a "precompetitive" or "procompetitive" nature, which the sponsors are free to exploit. However, determining the technological needs of the industry members, setting the research agenda, and helping meet the requirements of SME members were and will be among the most difficult challenges facing the IRAs.

Notes to Chapter 6

1. Natural Sciences and Engineering Research Council of Canada, *Ten Years to 2000: A Strategy Document* (Ottawa: Supply and Services Canada, 1989), p. 9.

2. Science Council of Canada, *Gearing Up for Global Markets* (Ottawa: Supply and Services Canada, 1988), p. 4.

3. National Science Foundation, *International Science and Technology Update: 1988* (NSF 89-307) (Washington, DC: 1988), p. 2.

4. Economic Council of Canada, "High-Tech Trade: Serious Weaknesses to Solve," *Au Courant,* 10, no. 2 (1989), 10.

5. J. N. H. Britton, and J. M. Gilmour, *The Weakest Link,* Background Study 43 (Ottawa: Science Council of Canada, 1978), p. 151; and Economic Council of Canada, *The Bottom Line: Technology, Trade and Income Growth* (Ottawa; ECC, 1983).

6. *The Urgent Investment: A Long Range Plan for the National Research Council of Canada,* Ottawa (1980), p. 68.

7. NSERC, *Ten Years to 2000,* p. 6.

8. ECC, *Bottom Line,* p. 82.

9. Ibid.

10. Ibid.

11. Science Council of Canada, *Technology Transfer: Government Laboratories to Manufacturing Industry,* Report No. 24, (Ottawa: Government of Canada 1975), pp. 46–48.

12. Ibid., p. 20.

13. Britton and Gilmour, *Weakest Link,* pp. 180–82.

14. Norman Eaton, "The Role of the Industry Associations: The Welding Institute Story," 1984 CRMA Annual Conference, Winnipeg, pp. 6–7.

15. Supplied by SPI.

16. Britton and Gilmour, *Weakest Link,* pp. 180–81.

17. *SPI Dimensions,* January–March 1984.

18. CSIRA Annual Report.

19. T. J. Allen, D. B. Hyman, and D. L. Pinckney, "Transferring Technology to the Small Manufacturing Firm: A Study of Technology Transfer in Three Countries," *Research Policy,* 12, no. 4 (August 1983), 209.

Chapter 7

Preparation of Technology Transfer Agents

Michael H. Bernhart

ROLE OF AGENTS IN TECHNOLOGY TRANSFER

The transfer of technology often, perhaps usually, involves human agents to assist in the transfer. Whereas technology transfer between two organizations with similar technical skills might be effected through the simple movement of a piece of machinery from one to the other, the need for human agents to assist the process probably increases when either of two conditions arises: (1) there is a technological gap between the sending and receiving organizations, and (2) the technology is not machine-embodied (for examples of machine-independent technologies, see Chapter 9, which discusses the transfer of social/organization technologies).

Skills Required

It would appear that the agent can perform a wide variety of functions, including the following:
  Prior to transfer the agent can

- Sell the technology.
- Negotiate the terms for its transfer.
- Establish the transfer mechanisms.
- Identify requirements for adapting the technology.

During transfer the agent can

- Prepare, adapt, or modify the technology.
- Prepare training materials.
- Install the technology.
- Provide formal and informal training.
- Oversee operations.

After transfer the agent may

- Evaluate the performance and suitability of the technology.
- Revise or re-adapt the technology.
- Monitor quality control.
- Monitor contract compliance.
- Monitor the supply of inputs.
- Identify innovations for back-transfer.

In all of these activities, the agent, if he or she is employed by an organization with a long view of the international market for technology, will be working to solve problems and maximize client satisfaction.

The tasks listed require a mixture of skills, only one of which is technical competence. In fact, technical skills may be much less important to managerial success than "relational" skills in many societies. (See Montgomery for a survey that documented this in nine African countries.)[1] This is worth remarking upon because it appears that the principal criterion for selection of technology transfer agents has been their presumed technical competence; little emphasis has been placed on relational skills.[2] This tendency is easy to understand when one considers the relative ease of determining technical ability as contrasted with the difficulties encountered in assessing relational skills. How, for example, do you quantify a person's capacity for empathy—or tolerance for ambiguity? A second reason why an organization may assign primacy to technical skills is that the presence of such skills does reduce the likelihood of immediate failure.

THE FAILURE OF U.S. EXPATRIATES

Granting that technical skill is a necessary requirement for the success of a technology transfer agent (TTA), it is not a sufficient one. It is now old news that relational skills will be an important determinant of whether the agent is ultimately successful. The evidence we have at hand is that the agent may *not* be successful. Hubbard reported that 33 percent of U.S. expatriates have to be recalled.[3] Mendenhall and Oddou put the recall rate at 20 percent but stated that many additional U.S. expatriates who are permitted to complete their assignments are quiet failures.[4] Hixon examined the literature and estimated that 25 percent of U.S. expatriates are failures and another 25 percent are marginal performers or hidden failures.[5]

These are discouraging figures, but they are consistent with findings that have been reported over the years. Anecdotal and survey evidence that has been collected to explain this high failure rate has fingered the inability of U.S. expatriates to adapt to the cultural environments in which they are placed. Representative of this research are the findings reported by Tung.[6] In a survey of 80 U.S. multinationals, the five reasons most often cited to explain the high failure rate of U.S. expatriates were a variety of adjustment problems suffered by the employee or the employee's dependents; lack of technical competence was ranked a weak sixth.

The literature on this topic has echoed a consistent theme for decades: U.S. expatriates do not perform as well as do their European and Japanese counterparts. The reason? U.S. firms do not provide as much predeparture preparation for their expatriate employees.[7] And despite the accumulating evidence that more training would reduce the probability of failure, it seems unlikely that U.S. firms will move quickly to provide such training. There are two reasons for this reluctance: (1) U.S. firms, it is averred, lack the international orientation that their foreign competitors possess. This may be, in part, owed to a system bias that will be slow to change, Adler reported that two-third of U.S. multinational CEOs had no foreign experience.[8] (2) Given the high mobility of U.S. managers between firms, a company may feel that a heavy investment in predeparture training for managers is ill-justified.

In sum, the outlook for the prospective U.S. expatriate is bleak. The chances that he or she will receive any predeparture preparation are small, 25 percent, according to Baker.[9] The likelihood that he or she will fail overseas is high—50 percent—and help is not on the way.

SELF-PREPARATION

The only available avenue for most expatriates is self-preparation. But urging the prospective sojourner to engage in contemplative study of the new culture during the hectic days prior to departure is a forlorn hope. It seems more likely that the expatriate will find both the motivation and opportunity for study of the culture after arrival. As noted later, perhaps only after arrival can the TTA study those things of greatest relevance to his or her assignment. Further, the task-specific knowledge needed is unlikely to be covered in a general predeparture course so the TTA, even if one of the fortunate few who attends a training course, will still bear the full burden of equipping him- or herself for the task.

The remainder of this chapter outlines the areas of information of most immediate relevance to the TTA, an approach for efficient acquisition of that knowledge and the skills necessary for successful completion of his or her assignment.

RELEVANT KNOWLEDGE

What, of the infinite display of culturally specific information, is most relevant to the success of the TTA? It is assumed that the technology transfer agent, unlike the expatriate manager, will have an assignment that is both closed-ended and short. It is further assumed that a specific task will have been spelled out in advance, or at least defined by the agent shortly after arrival.

History, Politics, and Art

Given the tight time constraints, the agent will have to be selective regarding which aspects of the host culture he or she will attempt to master. Probably sacrificed under these conditions will be study of the history, politics, or art of the society. Knowledge of these would enrich the overseas experience for the TTA, but they may not be *absolutely* essential for accomplishing the agent's task, and time might be better dedicated to acquiring task-relevant knowledge. Of course, exposure to the long history and proud heritage of other societies is a useful antidote to cultural chauvinism. However, the hope grows that as American pre-eminence in the world economy wanes, our cultural arrogance fades with it.

Social Etiquette

What of etiquette and protocol? The cross-cultural faux pas that provide the lead-in to articles on cultural misunderstanding sound deeply embarrassing and we might believe that they are the sole cause of the breakdown of negotiations or the collapse of a career. One American points his toe at or touches the head of a Thai associate; another hapless expatriate displays the sole of his shoe to a dismayed Arab host; a third mars an otherwise pleasant luncheon with his Japanese counterparts by the unseemly position of his chopsticks on the plate; and so on. Are the cultural waters always that perilous? Consultants and trainers hawking instant cultural sensitivity seem to suggest that they are; research results are mixed.

Collett ran an experiment with Arabs and Englishmen to investigate whether a patina of cultural refinement contributed to improved understanding and collaboration between members of the two cultures.[10] He divided his British research subjects into two groups. One group was given an article (drawn from Watson and Graves)[11] on Arab etiquette and protocol. These Britons learned that Arabs prefer to sit head-on and close during conversations, stare directly into the other's eyes as a sign of attentiveness, and so on. The other group of Britons read a brief treatise on Arab history. After the Arabs had a chance to converse with members of both groups of Englishmen, they were asked to rate them on a number of dimensions. The brief training in Arab etiquette paid off: the Arabs preferred the trained Englishmen; they would introduce them to their families, saw them as nicer, felt they understood the Arab better, would rather be friends with them, and would share living quarters with them. *However,* the Arab subjects were equally divided when it came to choosing with whom they would be willing to continue the conversation or to enter into business dealings; the untrained Britons fared no worse than the trained ones.

Why this evident indifference to social skills when it came to the task? One answer might be that the Arab subjects had learned to expect and tolerate non-Arab etiquette. Lindgren and Tebcherani[12] asked Arab and American students to fill out a questionnaire and then predict how members of the other culture might fill it out. The Arabs were uncanny in their ability to predict how Americans would answer; the Americans failed miserably at predicting Arab answers. It would appear that the Arabs in the study know us much better than we know them.

If there is a lesson to be drawn from these two studies, it may be that the sojourner can still function at the task level while failing at the social level. Three qualifications to this assertion would seem prudent: (1) the sojourner is interacting with hosts who have had some exposure to the sojourner's culture; (2) observation of etiquette is not paramount in the society (as it may be in Japan); and (3) the sojourner has not been in the host society for such a long time that his or her continued ignorance of local customs bespeaks indifference and disrespect. These caveats noted, however, in many, perhaps most, situations the naive American can probably muddle through, often amusing his or her hosts, occasionally embarrassing him- or herself, and reducing opportunities for social contact through social blunders. But he or she should still be able to get the job done.

Note that if the American wants to do better on the social score, little investment is required. Collett's trained subjects spent less than 30 minutes studying Arab etiquette and the popular literature on culture-specific etiquette is now extensive. A more insidious danger than ignorance of local etiquette would seem to be the inability of the American to respond equably to breaches of American etiquette. Any hint of displeasure, no matter how subtle, that the American might exhibit in response to the normal exercise of local customs would be more offensive than failure to adopt those customs.

Task Relevant Behavior

If it is impractical for the short-term TTA to learn the history, art, and politics of the host culture and it is likely that he or she can survive with only scant knowledge of local etiquette (although such knowledge can be easily acquired). What cultural knowledge is absolutely indispensable? The answer is the same at home as it is abroad: the agent has to know how to get things done in an organization—how decisions are made and implemented.

An important assumption in the following is that the TTA is a bright person and capable of solving problems once he or she knows what those problems are; an interest in and capacity for problem solution seems consistent with the technical skills for which the TTA was selected. The real trick for the transfer agent is to divine the issues raised by the adoption of the technology transferred. Whereas the specific issues will vary with the responsibilities of the agent and the differences between the sending and receiving organizations, the following would seem to be fairly common questions that will have to be addressed:

What behaviors have to change?
  What old behaviors have to be extinguished?
  What old behaviors have to be modified?
What new behaviors have to be learned?
How are changes accomplished?
  How do people learn?
  How are decisions made?
  What supporting institutions exist that will facilitate or impede the adoption of new or modification of old behaviors?

Supporting institutions. Starting with the last mentioned as the potentially most troublesome, here are two examples that exemplify the problem and opportunity.

> A century ago British colonizers, recognizing the dietary limits imposed upon the Maoris by lack of modern hunting equipment, provided the tribesmen with firearms. The innovation worked; more game was killed. But at the time of the technology transfer, the Maoris settled inter- and intratribal conflicts by resort to arms—typically clubs, spears, rocks, etc. That this approach to conflict resolution was dysfunctional in the age of rifles goes without saying;

within a few years the tribes were decimated before new institutions arose to facilitate the settling of disputes.[13]

Bliss reports a happier outcome when the wheel was introduced to Papago Indians. This would seem to be a tough innovation to absorb for it ". . . displaced some parts of the technology and established new techniques and specialties . . . resulted in important shifts in the division of labor, had far-reaching effects on the economy, became for a period a strong factor for greater community solidarity, and influenced the relations of the Papagos with surrounding peoples. (p. 32)" Despite the pervasive influence of this innovation, all of these changes were accomplished in an orderly, rapid, and nondisruptive fashion.[14]

Examination of these two examples reveals some interesting contrasts. It would be easy, but incomplete, to conclude that the sole explanation for the magnitude of the differences between the two examples resides in the destructiveness of weapons and benign utility of the wheel; the differences were more complex. Reintegration of the institutions affected by the wheel was relatively direct. The Papagos were already highly mobile, depending upon pack horses; Indian agents introduced the accompanying innovations (harness making, smithing) that facilitated full adoption. The most serious obstacle was lack of roads, but the tradition of communal labor on projects was long established through hunts and food gathering. All the attendant innovations were in existence, to some extent known by the Papagos, encouraged by agents, and consistent with or already part of existing institutions.

Not so with Maoris and firearms. Although the firearms were demonstrably useful in hunting, incessant low-level warfare was an institution broadly imbedded in the culture. It served the multiple functions of dispute resolution, territorial maintenance, ecological balance with game, and rites of passage to manhood. In contrast to the Papago example, there were no immediately available alternatives, thus requiring the invention of new institutions to perform the functions of passage, conflict resolution, etc.

Questions for the TTA. These two examples were chosen to dramatize the potential for trouble as well as the fact that ease of initial adoption (the Maoris readily embraced firearms) does not signal successful integration of the technology. The TTA has to ask whether the new technology

1. Distorts social patterns? Trist's and Bamforth's classic study of the failure of new technology in coal mining demonstrated how important social relationships can be in the workplace.[15]
2. Upsets status differentials? Does the new technology change the loci of authority in the organization? Are quality control inspectors suddenly more influential than traditional supervisors?

3. Invalidates the reward system? Does the piece-rate system no longer apply, forcing the organization to pay hourly wages?
4. Requires complementing skills and technologies? As existed in the case of the Papagos.

The TTA has to compare the status quo with the demands made in these four areas by the new technology. It is not difficult to find answers to these four questions, if one bothers to ask them.

Accomplishing change. A second nettlesome area for the TTA is determining how change is accomplished within an organization. Almost certainly two questions will arise: (1) how are decisions made so that change may be legitimated? and (2) how do people learn?

We are now well informed on the different decision-making style of the Japanese. Less well documented, however, is the decision-making style in Indonesia—apparently consensus is reached in (interminable to a Westerner) meetings attended by large groups of people. Or in Hong Kong where senior executives appear to occupy ceremonial rather than decision-making roles.[16]

Questions for the TTA. The TTA will have to know where to go for approval of a training program, or simply for authorization to dispatch an urgent request for materials. Before seeking that authorization, he or she will need to know the following:

1. For a given issue, who may initiate a request for a decision? Often it is not the TTA.
2. What information is expected in support of such a request?
3. How are these requests communicated?
4. Who is involved in the decision-making process?
5. What are their roles? Some must be consulted (can voice an opinion but cannot veto), some must approve (can veto), and someone, at some point, must authorize. This is a tricky area; it is easy for the outsider to confuse approval by an executive with authorization by the organization and to move forward prematurely.
6. What is the sequence of approving and authorizing?

As with the preceding set of questions, these are researchable issues (and they are questions that often go unanswered in the agent's home organization) if the small effort required is invested.

How do people learn? Technology transfer, almost by definition, means that the recipients will have to acquire new skills, presumably through training. The most cursory examination of educational practices in the world indicates that societies differ with respect to how knowledge is transmitted.

The learner may be an active or passive participant. The learner may be expected to challenge the material, or to remain silent. Written material may play a large or small role. And so on. Beyond educational philosophies, experience with different educational technologies may be great; Guthrie and coworkers cite the example of Vietnamese technicians who were unfamiliar with blueprints.[17]

Questions for the TTA include:

1. How are instructions communicated? In writing? As verbal commands? As requests?
2. How is understanding demonstrated? Does silence mean comprehension?
3. Who is a credible teacher?
4. Are the learners literate?
5. Do they comprehend the language of instruction?
   The formal language—English, Sinhalese, etc?
   The argot of the profession?

There may be some universally superior approaches to education. For example, active learning may be better than passive. Such superiority notwithstanding, if a novel pedagogical approach is introduced, learners will first have to master it, and its very novelty will, at least initially, be a distraction. Even if the TTA is convinced of the rightness of his or her educational approach, it will still be useful to determine if that approach is at variance with the experience of the learners.

What behaviors have to change? The final challenge for the TTA is to determine which behaviors have to change. Which behaviors that were functional with the old technology no longer serve? Which old behaviors have to be modified? And what new behaviors have to be instituted—behaviors for which there is not precedent in the experience of the recipients of the technology?

These behaviors fall into three broad categories: technical skills, attitudes, and social behaviors. Easiest to transfer are the *technical skills* required by the new technology. The TTA might expect, and be expected by the hosts, to devote most attention to these. First, the TTA's credibility in this area is presumed. Second, everyone anticipated that the new technology would involve acquisition of new technical skills and they are prepared for this kind of change. Third, traditional training methods are well suited to imparting these skills. And finally, progress on this front is easy to verify. Acquisition of such skills may be, however, only a small part of the total changes that need to be made.

More difficult are changes in *attitudes* that the new technology may require. For example, stories abound concerning misperceptions of the role of human operators in highly automated operations. Representative of this genre is the story of the hapless man employed to monitor the warning system on generators installed on the Upper Volta. If a light comes on, he was told, shut down the system. He expected continuous interaction with the machinery, as had been his previous experience. After watching for a light for several days, he concluded that no interaction was necessary and started to nap on the job with calamitous results.

Another difficult, and perhaps resistant, area is that of changes required in *social behavior*. Will people be willing to work in greater isolation than before? Or does the technology increase the noise level so that conversation becomes difficult? Is punctuality now more important? Determining which social behaviors will have to change will challenge the transfer agent on two fronts. First, the simple identification of these needed changes requires a great deal of insight into the workings of the host culture. It seems highly unlikely that any general predeparture training course will deal with issues such as these. Second, the TTA will have to be careful that his or her cultural chauvinism is not aroused by the mismatch between the host culture and imported technology. The agent will have to recognize the mismatch and then attempt to modify either the social behaviors, or the technology, or both to reduce the gap.

Where the cultural distance between the sending and receiving societies is the greatest, these changes may present the thorniest problem for the TTA. His or her perceived legitimacy to encourage changes in social behaviors may be low; the incompatibility between the technology and the host culture may not be evident to many host managers; and the methods for changing social behaviors are uncertain in their efficacy. In those instances where the technology can be modified to reduce the gap, the TTA may find this a far easier avenue.

The range of social behaviors where the technology might be incompatible with the host culture is vast; can any core areas be identified? The answer, unhappily, is that only the most general questions can be proposed.

Questions for the TTA include

1. What social interaction did the old system permit and does that differ from the new system? Will people be able to talk? Choose their own workmates? Stop briefly when they want to for horseplay or conversation?
2. What of status differentials? Do supervisors retain the same perogatives? Can they assign workers? Control the supply of inputs? Who would a worker go to under the new system to solve a problem?
3. What control did and will the worker have over his or her work? Will

workers be able to vary the pace? Will they be able to vary the tasks they do or the order in which they are done? Can they relax when they want to?

4. Is the reward structure likely to change? Will the relationship between effort and rewards be clear? Were there intrinsic rewards under the old technology that will be absent under the new? Will the worker see an end-product?

These are difficult questions. Considerable insight will be required to come up with the appropriate questions and only keen observational skills will lead to correct answers. In reviewing these questions it seems clear that predeparture courses will not address issues this specific to the TTA's task (which is not to argue that such courses should be avoided; they can do great good and are injurious only when the participant smugly assumes that his or her cultural education is complete). One reassuring note: the TTA need not change social behavior; he or she need only detect where the technology is incompatible with host behaviors. His or her responsibilities at that point are to modify the technology to reduce the variance and to inform host managers and supervisors of problem areas. It ultimately will fall upon host management to change the perceptions, behaviors, and attitudes of the local workforce so that the technology is successfully utilized. They cannot, however, be expected to anticipate the demands that an unknown technology will place upon the established social system. The TTA, who will know well (we hope) the new technology and who will be able to observe the system it replaces, is better positioned to identify the major changes that will have to be made.

STUDYING CULTURE

The preceding section may indicate that the TTA will have to be as much a social scientist as a technician. That is exactly correct. This claim is made without apology; the TTA *should* be a social scientist, who examines social phenomena through observation, questioning, and controlled experiments. The TTA has been using these social research skills since birth to understand how to function in the world; he or she will simply have to do it again. Not coincidentally, Schild found that sojourners relied on the same three data collection methods, observation, questioning, and experimentation, in their efforts to master a new culture.[18] But too few people do this well. It is discouraging to read the findings of Byrnes,[19] who found in his study of U.S. technical advisors that they lacked depth and range in asking questions and studying the host culture. Less than one-fourth of those studied demonstrated insight into the decision-making process of their hosts and just better than half had read published material on the host country.

The preceding section discusses the types of question that the agent will need to address. This section describes an approach for obtaining answers to those questions—how to research the local environment. To illustrate the broad range of behavior that may be studied, examples of published research and projects that university students in Atlanta have undertaken are presented.

A Bad Example

We are all familiar with the travelers who return home laden with knick-knacks and unending tales of foreign thievery, boorishness, and deceit. Examination of these stories might reveal that the conclusions rest on flimsy evidence: the research samples were unrepresentative (members of the tourist service industry) and small (one cab driver); the method of data recording was unreliable (the husband and wife typically disagree on the details); and the effect the researcher/traveler had on the outcome is rarely considered.

A Good Example

Intrigued by the popular American folk wisdom concerning the dishonesty of foreigners, Roy Feldman determined to put the question to the test.[20] It had occurred to Feldman that this sweeping generalization might be the product of methodological errors. He employed a native and a "tourist" to conduct four experiments in Boston, Athens, and Paris. In each experiment the behavior of the two researchers was carefully controlled and data were methodically recorded. In one experiment the researchers asked directions to a well-known landmark. In the second, passers-by were asked if they had dropped some money on the sidewalk (money planted by the researcher). In the third, cashiers overrefunded money to purchasers, and in the fourth the fares charged by cabbies to the native and tourist were compared. No need to go into the melancholy details; Boston did not fare well. Bostonians tended to be less helpful and honest with foreigners than with a fellow American. Parisians and Athenians tended to treat foreigners as well or better than their own countrymen. (The stand-out exceptions were Paris taxi drivers who demonstrated a fine and roguish ability to extract premium fares from foreigners.)

This study is illuminating, not only because it shatters the old myth of the categorical unscrupulousness of foreigners, but because it also demonstrates how easy, even fun, cross-cultural research can be.

Research Through Observation

Observation of the culture is the easiest learning method. It is performed almost effortlessly and does not depend upon the host's active contribution to our learning.

The first task when using this research method is to select appropriate subjects for observation. Good subjects are those who are doing things that the TTA will need to do; suspect subjects are those that the TTA feels "at home" with. It is a simple matter to note how the subjects behave. They stand when another person enters the room; they shake hands frequently; they never smile when another is talking; and so on. If the TTA is attentive, he or she should have some idea, before leaving the arrival airport, how people greet each other, what attitude one takes with customs officials (and perhaps other petty bureaucrats), whether cash is openly displayed, and so on. These may not be momentous learnings, but they illustrate the ubiquitousness of cultural information. Three obvious forms of observation-based research are watching, listening, and reading.

Watching. Social science has made frequent use of watching in cross-cultural research. Berkowitz[21] had his researchers watch pedestrians in Italy, West Germany, England, Sweden, and the United States; they recorded social interactions—who talked with whom, who touched, who was abroad in the streets, etc. Others have followed in Berkowitz's footsteps by directly observing subjects, and Weston LaBarre recommended the movies as an additional source of data on gestures and modes of expression.[22] To exemplify the range of behavior that may be observed through watching, consider the conclusions students arrived at in simple research projects:

- Mid- and large-size passenger cars are more likely to exceed the speed limit than are small cars.
- Men drivers are less likely to heed stop signs or use turn signals than are women.
- Married female employees select more risque greeting cards than do housewives.
- Theology students are heavier smokers than are business students.
- Black female students are more likely to carry a briefcase than are white female students.
- Women and male white collar employees walk faster than other pedestrians.

These results are probably credible for the year and location of the studies. The students who conducted them were systematic and thorough, yet, true to the traditions of university students everywhere, they did not invest heavily of their own time. Systematic observation of members of a culture is simple, reliable, instructive, and often an antidote to boredom.

Listening. If one is fluent in the native tongue, an especially rich source of data is available. Moore, in an early study, eavesdropped on early-evening strollers, and Landis replicated the experiment overseas.[23] Interesting differences appeared concerning the topics of conversation and how those topics changed as the gender composition of the group changed.

Students have eavesdropped to catalogue the relative frequency of off-color words, to determine who is least willing to supply a first name in business dealings, and how subordinates address their superiors in different settings. Again, the methods were simple but adequate. Samples were representative of the population under study and data were recorded at the time of observation.

Reading. Examination of archival material is a natural and it is easy to be systematic. In a novel use of printed material, Janes used the society page of local newspapers as a guide to community structure.[24] No one person has prompted more interest in cross-cultural research than did Adolph Hitler. Frequent use of written materials has been made to examine the values at work in Nazi Germany; McGranahan and Wayne (1948) analyzed plays; Lasswell (1941) looked to newspaper reporting; and Sebold (1962) found significant information in song books.[25] Students with access to company records studied the following:

- The racial composition of neighborhoods and how prices and foreclosure rates varied.
- Sick leave taken by job category (nonprofessionals tended to be sick on Mondays; professionals were sick on Fridays).
- How academic performance relates to the first letter of a student's last name (it is better to come earlier in the alphabet).

Direct Communication

Asking questions is a natural; however, there are two issues that have to be addressed first by the sojourner. Is it acceptable to ask questions—about anything? And how do you interpret the response?

Can you ask anything? Research would seem to indicate that, whereas national sensitivities do vary, taboo subjects are pretty much the same around the globe; however, in some societies the threshhold is reached sooner than in others. Plogg compared the willingness of Germans and Americans to disclose personal information; Melikian studied the self-disclosure habits of Middle-Easterners; and Jourard compared British and American women.[26] In all three studies Americans were the most self-revealing group; we talked most freely about political views, habits, and interests, personal finances, morality and sex, and so on. However, the rank ordering of items was the same across the studies; a delicate subject in the United States was a delicate subject in the other countries studied and a freely discussed topic in our society was likely to be open for discussion abroad. In sum, it would appear that the TTA should feel free to ask any of the questions listed above.

How are answers to be interpreted? Does yes mean yes—or something else? Here there is no substitute for a reliable informant—or perhaps several informants—who can provide a cultural lexicon. The dangers of misinterpreting information are real and subtle. Alexander and coworkers[27] found that Indian and Spanish managers tended to provide answers about personal values that were consistent with their perceptions of Western values; when these managers were asked to describe the value system of their countrymen, they painted a more traditional picture.

As direct communication has been a mainstay of cross-cultural research, it is unnecessary to cite extensive examples of its application. A few references to the issues addressed by naive researchers, students, may encourage the sojourner to undertake more aggressive and systematic questioning of his or her own:

- Do men and women differ in their use of credit cards?
- What attributes do personnel officers look for in hiring interviews?
- Do foreign students return home after college?
- How many women sleep in the nude?
- How do the characteristics of Volvo and BMW owners differ?

Experimentation

The first two methods of data gathering are fairly safe in that they permit the TTA to maintain some distance between him- or herself and the objects of study. The third method, experimentation or direct participation, allows no refuge. With this approach the TTA experiments with different behaviors and gauges the response. This means exposing oneself to possible rebukes and failures, yet remaining sufficiently detached so that the results can be studied dispassionately.

Because the response to an experiment may be nonverbal, it may initially be difficult for the TTA to interpret the meaning of results. The suggested solution is to restrict experiments to situations where you want a clear behavioral outcome. For example, you may want greater respect from your host colleagues (a difficult issue to research through questions or observation). Respect is difficult to measure in the abstract, so you will need to gauge success through other measures such as response time to your requests or attendance and punctuality at meetings you call. To obtain more respect, you might vary your clothing, working hours, medium of office communication, etc.

Feldman's study of honesty[28] is a minor classic in that it illustrates the ease with which major issues may be addressed with simple methods. Students have also had no trouble in devising home-grown experiments. One extended his left hand in greeting (blacks took it; whites were nonplussed); another dropped pencils in crowded elevators around town to compare willingness to help (the least help came around law offices); and another coached her niece to approach strangers and declare that she was lost (everybody offered protection and assistance to the child).

SUMMARY

This approach may sound very academic and ponderous. It may be, but it is also systematic. As a personal aside, I have failed on enough assignments to know that there is a wrong way as well as a right way to do things. The right way is to get into the routine of carefully cataloguing what I see, hear, and experience. The research quickly becomes a pleasant diversion.

It would be pleasant if we all had the time and wits to master every situation. The technology transfer agent is keenly aware that this is not possible. Typically he or she gets little advance warning of an assignment and little time in which to complete it. Under these conditions we cannot expect these men and women to master all aspects of a society (we could, however, expect a great deal more of managers assigned for longer periods). The best we might ask for is that they stay out of trouble and learn enough to complete the job. And there is plenty to learn. They have to learn how to learn—how to interpret the information they receive. They have to learn what changes are required by the technology they are bringing with them—the changes in technical knowledge, attitudes, and perceptions, and social behaviors. They have to learn how the organization works—how decisions are made and implemented. And they have to learn how to communicate knowledge to their hosts. Lamentably, little of this information is found in preparatory courses—where such courses are available—and the burden falls on the TTA to work out the answers for him- or herself. To work through all the questions that are central to the success of the TTA requires considerable self-discipline and efficiency. Like everyone else, the TTA is trying to figure out the system; the crux of the problem is being systematic.

Notes to Chapter 7

1. John D. Montgomery, "Comparative administration: theory and experience." *International Journal of Public Administration*, 12, (3) (1989), 501–12.

2. E. L. Miller, "The selection decision for an international assignment: a study of the decision maker's behavior." *Journal of International Business Studies*, 3 (1972), 49–65.

3. Ghislaine Hubbard, "How to combat culture shock." *Management Today*, September (1986), 62–65.

4. Mark Mendenhall, and Gary Oddou, "The overseas assignment; a practical look." *Business Horizons*, 31 (5) (1988), 78–84.

5. Allen L. Hixon, "Why corporations make haphazard overseas staffing decisions." *Personnel Administrator*, 31 (3) (1986), 91–94.

6. R. L. Tung, "Selection and training of US, European, and Japanese multinationals." *California Management Review*, 25 (1) (1982), 57–71.

7. Chris Brewster, "Managing expatriates." *International Journal of Manpower*, 9 (2) (1988), 17–20; and R. L. Tung, "Expatriate assignments: enhancing success and minimizing failure." *Academy of Management Executive*, 1 (2) (1987), 117–125.

8. Nancy Adler, *The International Dimensions of Organizational Behavior* (Boston: PWS Kent, 1986).

9. James C. Baker, "Foreign language and pre-departure training in US multinational firms." *Personnel Administrator*, 29 (7) (1984), 68–72.

10. P. Collett, "Training Englishmen in the non-verbal behavior of Arabs." *International Journal of Psychology*, 6 (3) (1971), 209–215.

11. O. M. Watson, and T. D. Graves, "Quantitative research in proxemic behavior." *American Anthropologist*, 68 (1966), 971–985.

12. H. C. Lindgren, and A. Tebcherani, "American and Arab auto- and heterostereotypes: a cross-cultural study of empathy." *Journal of Cross-Cultural Psychology*, 2 (2) (1971), 173–180.

13. G. Lizitsky, *Four Ways of Being Human* (New York: Viking Press, 1956).

14. Wesley L. Bliss, "In the wake of the wheel: Introduction of the wagon to the Papago Indians of southern Arizona. In Edward H. Spicer, ed: *Human Problems of Technological Change*, (New York: Russell Sage Foundation, 1952).

15. E. L. Trist, and K. W. Bamforth, "Some social and psychological consequences of the longwall method of coal getting." *Human Relations*, (February 1951), pp. 3–38.

16. W. A. Evans, D. Sculli, and W. S. L. Yau, "Cross-cultural factors in the identification of managerial potential." *Journal of General Management*, 13 (1) (1987), 52–59.

17. G. M. Guthrie, H. W. Sinaiko, and R. Brislin, "Non-verbal abilities of Americans and Vietnamese." *Journal of Social Psychology*, 84 (2) (1971), 183–190.

18. E. O. Schild, "The foreign student, as a stranger, learning the norms of the host culture." *Journal of Social Issues*, 18 (1) (1962), 41–54.

19. F. C. Byrnes, *Americans in Technical Assistance* (New York: Praeger, 1965).

20. Roy E. Feldman, "Response to compatriot and foreigner who seek assistance." *Journal of Personality and Social Psychology,* 4 (1) (1968), 202–214.

21. William R. Berkowitz, "A cross-national comparison of some social patterns of urban pedestrians." *Journal of Cross-Cultural Psychology,* 2 (2) (1971), 129–144.

22. W. LaBarre, "Paralinguistics, kinesics, and cultural anthropology." In Thomas Sebeak, ed.: *Approaches to Semiotics,* (The Hague: Mouton, 1964).

23. C. Landis, "National differences in conversation." *Journal of Abnormal and Social Psychology,* 21 (1927), 354–57; and H. T. Moore, "Further data concerning sex differences." *Journal of Abnormal and Social Psychology,* 17 (1922), 210–214.

24. R. W. Janes, "A technique for describing community structure through newspaper analysis." *Social Forces,* 37 (1958), 102–109.

25. H. D. Lasswell, "The world attention survey." *Public Opinion Quarterly,* 5 (1941), 456–462; S. M. Jourard, "Self-disclosure patterns of British and American college females." *Journal of Social Psychology,* 54 (1961), 315–320; D. McGranahan, and I. Wayne, "German and American traits reflected in popular drama." *Human Relations,* 1 (1948), 429–455; and, H. Sebold, "Studying national character through comparative content analysis." *Social Forces,* 40 (1962), 318–322.

26. L. H. Melikian, "Self-disclosure among university students in the middleeast." *Journal of Social Psychology,* 57 (1962), 257–263; and, S. C. Plogg, "The disclosure of self in the United States and Germany." *Journal of Social Psychology,* 65 (1965), 193–203.

27. R. A. Alexander, G. V. Barrett, B. M. Bass, and E. C. Ryterband, "Empathy, projection and negation in seven countries." Technical Report #35, Management Research Center, University of Rochester, 1970.

28. Feldman, *op cit.*

Chapter 8

Transferring Soft Technology

Bruce Morgan

International technology transfer is often remarked on and sometimes described, but it has seldom been analyzed as a process. This chapter looks at the transfer process in the context of "soft" technologies, such as management or planning.

Much of what this chapter contains will strike practitioners of soft technology transfer as simple common sense; at least that is the aim. Most soft technologies, however labeled, are not arcane systems or scientific practices. They are simply rational organization, discipline, and routine psychology. Further, the transfer process itself is, necessarily, a pragmatic, common-sense exercise.

In working with developing countries trying to adopt Western technologies, I have followed what little literature there is on technology transfer, but I have not seen much useful analysis of the transfer process as such. (However, there is considerable literature on other, related topics, and I refer to some of those subjects in passing.) The judgments in this chapter are just that; personal opinions with no specific support from empirical data or academic studies. This series of propositions is set out in the hope that more scholarly readers will devote greater attention to this important but neglected are of technology transfer.

SOFT TECHNOLOGIES—A DEFINITION

Chapter 1 discusses the meaning of "technology" when used in the context of international technology transfer. Although there is no clear dividing line between hard and soft technologies, the basic difference is that hard technologies are tangible or well defined, whereas soft technologies are not.

Before describing the difficulties with soft technology transfer and how those problems can be overcome, a working definition of "soft" technology is necessary.

I define soft technologies as learned behaviors: skills and methodologies that do not have fixed technical meanings or formulas and may even be subject to ambiguity or controversy in their original environment. Examples include management (whether one considers it a science or an art), manpower development and related topics, organizational development and its related disciplines, and many innovative practices in any field, whether it be health, agriculture, or social welfare.

Soft technologies, by this definition, are simply human skills or practices, not "things" to be transplanted. This distinction is arbitrary and could lead to confusion (computer software is hard technology by this definition, for example). However, making the distinction is important in order to analyze why transferring such technologies is difficult.

In most soft technology transfers, the new user must change some existing behaviors. This is the major element making successful adaptation difficult. (The words "transfer," "innovation," and "adaptation" are used here relatively interchangeably. "Transfer," however, carries a connotation of logistics and the technology provider's role; "adaptation" suggests emphasis on the nature of the change and the user's environment.) Another factor is that these skills usually cannot be directly or easily demonstrated. (One can teach the use of a lathe by using it, but even if demonstrating a public health practice, for example, is possible, the improved result usually is not evident.)

If the technology has these two characteristics—it requires behavioral change and cannot be adequately demonstrated—then a soft technology is involved and greater care must be taken in designing and implementing the transfer process.

Even a major technology change can be managed as though it were a "hardware" transfer if it does not require a significant change in behaviors by the users. (An example would be changing to touch-tone telephones.) In contrast, a relatively modest innovation that *does* require changing behaviors (such as a new agricultural practice) has to be viewed primarily as a soft technology transfer.

The behavioral nature of soft technologies makes them heavily dependent on the individual talents of those responsible for the transfer and the attitudes and working styles of those who are to use the technology. This reliance on personal characteristics is what makes their transfer generally more difficult than the transfer of "hard" technologies.

Soft technologies are often lumped together under the rubric of "know-how." This umbrella acknowledges that these technologies are process-based and very often have much more to do with pragmatic experience than with concrete science. For example, transferring a planning methodology, adapting it to differing cultural and social norms, and then tinkering to make it work in the new situation becomes a dynamic, practically experimental endeavor. Introducing a new manufacturing process, in contrast, deals with a more fixed technological base, although many of the human dynamics may be similar.

TECHNOLOGY INNOVATIONS: COMMON ERRORS

Most problems presented by international technology transfers are also present in domestic technology innovations. By the same token, many of the special problems of soft technology transfer arise in most hard technology transfers as well, as most hard technologies require new skills and behaviors by the users. Indeed, many of the horror stories from foreign assistance programs are cases where the participants thought the process was "just" a hardware transfer. (The recent history of improvements in office technologies offers any number of examples.) For this reason, much of this discussion should be relevant to the transfer of hard technologies as well.

Before moving on to the difficulties peculiar to adapting soft technologies abroad, let us note common mistakes made in the innovation or transfer of *any* technology, hard or soft. These are pitfalls to be avoided in applying new technologies, whether home-grown or imported. International transfers of soft technologies, particularly to less developed countries, simply make such problems more acute.

This is not an exhaustive list, but many failures of technology innovations can be attributed to such errors as

- Overeagerness. Although usually laid to the door of the inexperienced person providing technical assistance, recipients can also be guilty of moving with more zest than prudence. They have heard about or seen the technology but fail to fully understand its real cost or the difficulties in making it work. Whoever is to blame, rushing into a new technology without adequate preparation is nearly always fatal to its success.
- Underestimating the difficulty of innovation. Glossing over the necessary underpinnings of a new technology is a related and, unfortunately, very common cause of technological failure.
- Failing to "sell" the change in technology. For any innovation to work, those who will actually implement the technology must know exactly what resources or training are required of them, and, more important, how they will benefit from the change. (For many soft technologies,

this subject is covered very well in the literature on management and organizational development.)
- Technological overkill. This is the mistake of adopting a technology that is beyond what the environment can support, or is more sophisticated or complex than the situation requires. The tendency to want the best, prettiest, or most impressive toy on the block adds to the burden of making any technology work in a new environment by placing too much emphasis on the technology itself rather than the results it is supposed to accomplish. Technology is never an end in itself, but only the means to achieve some larger purpose.

> We once studied the internal document system of a developing country ministry. What it had was basically sound, though there were some serious bottlenecks. Our recommendations suggested how those could be eliminated, as well as how the system could gradually evolve into a more modern system.
>
> Some ministry officials with U.S. experience were disappointed. The basic filing system was from the era of the British raj, based on a "box-and-post" technology. These officials assumed we would share their jaundiced view of such antiquity and recommend replacement with an automated U.S. filing system. Apart from illustrating the tendency to technological overkill, this was also an example of the client wanting a *result,* but some observers expecting a *product.*

- Organizational inertia. As noted in the literature on organizational development, change is resisted, actively or tacitly, by some elements of any organization. The *causes* of inertia, how it manifests itself, and where are important elements in planning technological innovations.
- Overlooking support systems. Many projects that seem to involve only a specific hard technology fail because other technologies and support systems that should have accompanied the hard technology were not considered. Nearly all technologies require supporting mechanisms that may or may not be obvious at first glance.

> Foreign aid agencies have built a lot of sugar mills in the developing world but often failed to pay sufficient attention to the agricultural requirements (sugar cane or beets) of the mills, the necessary transport systems, and the requisite marketing and distribution networks. This failure has typically resulted in under-utilization or closure of the mill, which then becomes an embarrassing memorial to those responsible for what was to have been a boon to the community.

All of these problems, whether arising in a domestic innovation or connected with an international technology transfer, are well known and emphasized by the existing literature; practitioners are generally well versed in avoiding them.

OBSTACLES TO SOFT TECHNOLOGY TRANSFER

Apart from these general sources of error, there are other obstacles to successful transfer that are primarily caused by the nature of soft technology. The most troublesome of these obstacles fall into four rough categories: inappropriate assumptions, cultural differences, communications, and the individual behavior of those involved in the transfer process. These clusters overlap, however, so a particular obstacle or barrier often can be analyzed from more than one perspective.

Assumptions

Every technology is based on assumptions about what it needs to work properly and about the environment in which it can be successfully used. Most assumptions about soft technologies are implicit rather than express; they are unstated, taken for granted, or simply not recognized as requirements. Examples might include: the expectation of open, direct communications within groups or organizations (rather than formal arrangements that protect personal and professional privacies), a belief in change as a desirable possibility in human affairs (as opposed to a loyalty for traditional patterns of behavior), or an assumption of deductive rather than inductive, or even intuitive, analysis.

Although the assumptions may or may not be entirely sound in the original environment, if the technology works one may presume they are correct, or nearly so, *in that environment*. When transferring that technology, then, one should examine the receiving environment to assure that the required assumptions apply there as well. When the necessary assumptions do *not* hold true in the receiving environment (and very frequently they do not), the application of the technology will almost inevitably fail.

Mistakes about assumptions are hard to anticipate in practice; it takes a great deal of introspection to identify and examine the fundamental assumptions upon which a technology is based. In the soft technologies, assumptions often relate to expected individual and group behavior. Having grown up with our own patterns of behavior, we are often not aware that there are other ways to act or that our preferred ways are not universal.

There is an additional problem for those from stable, developed societies. Having advanced, at least materially, we have forgotten a lot about the variables that must be taken into account to make a technology, any technology, work in a new environment.

Examining the assumptions that were actually made in the development of a soft technology is often not a very productive exercise. Although explicit assumptions perhaps *should* have been made in developing the technology, in most cases they were not because much of the foundation on which such technologies are based is taken for granted. (Many of the examples used later in this chapter illustrate this point as well.)

A better approach is to consider what assumptions *must* be implicit in the technology. For example, in planning, the implicit assumptions have to be that: goals can be set and achieved; taking the initiative through objective planning, rather than simply reacting to events as they occur, is desirable and worth the investment; and setting and achieving goal X will, in fact, have result Y. Although this seems pretty obvious, each of these assumptions, particularly in a developing nation environment, can be questionable in a given situation and should therefore be carefully analyzed.

To use a specific example, let us say that a country's educational authorities plan to strengthen the faculty and curricula of Provincial College. Their expressed goals are to gain higher accreditation and respect for its bachelors program, after which they want to assemble the resources required to grant masters and later doctoral degrees. This is all to occur over a 20-year time period.

The implicit assumptions are that there will be a sufficient, qualified student population to support these goals; the required faculty and curricula will be available, or can be assembled; graduates at these various levels will be able to find employment appropriate to their training and commensurate with their investment; and the requisite funding will be available at each stage of the project. Once these assumptions are made explicit, it is fairly easy to examine their validity. Each needs scrutiny *and* continued attention to ensure that they remain operative.

Mistakes about assumptions are just as frequent when the transfer is made by a member of the receiving environment who happens to be familiar with the new technology. In fact, many failed attempts at soft technology transfer have been inflicted upon developing nations by their own citizens. The only difference is that the mistake is more likely to be about hidden but essential assumptions required by the technology rather than about whether the assumptions hold true in the receiving environment.

> In Egypt the government once mandated that its ministries use "Management By Objectives." This was when MBO was first in fashion among practitioners of scientific management in North America. Whatever the merits of MBO in North America (the more rigid versions have largely passed from the scene), its application failed in Egypt.

The reason, one could argue, is that MBO requires such alleged Western institutional characteristics as open and candid communications across bureaucratic layers, "feedback," and a very linear approach to planning (and then achieving) institutional goals. To a very large extent, these were simply not present in the Egyptian bureaucracies.

Assumptions are made on both sides of the transfer, and one is the recipient's choice of technology. The preferred situation, of course, is for the recipient and the expert to work together before the technology is specified; the recipient analyzing what is needed and can be supported, and the expert presenting the alternatives, requirements, and likely outcomes. This process also helps ensure that both sides understand the challenge and anticipate possible pitfalls.

In an effort to avoid the arrogance often associated with aid-givers, the Peace Corps in Latin America in its early years had a policy of responding only to "felt needs"—the agency avoided making judgments about what a community needed and instead tried to find out what the community itself believed was most needed.

Citizens of developing countries have well-developed political instincts, however, and quickly learned to provide answers that would please the Peace Corps leadership and Congress, its funding source. The most pressing "felt need," of course, was to ask for whatever was most likely to be provided.

This issue of who should choose the technology to be applied is sensitive, but simply meeting a request for a specific technology without considering the premises required for success and comparing alternate choices sets the stage for later recriminations if the technology fails.

Concern for this aspect of technology choice led to the concept of "appropriate technology." Not surprisingly, this was seen by Third World recipients as patronizing or exclusionary. The technology provider was telling them what they could have rather than providing what they wanted. The fact remains, however, that the most sophisticated technology (which is usually what is requested) is also typically the most expensive and not necessarily the best suited to a given situation, particularly in a developing country.

In examining the technology's assumptions, a parallel consideration is looking for what may be flaws in the technology itself, flaws that may not be apparent in domestic use but will show up in the new environment. When the developing country looks to experts abroad for assistance with soft technology, those responsible for providing it should consider whether the experience in the home country really warrants what is being claimed about its use in the new environment. (The tendency of the developing country recipient is to believe that these technologies are "sciences," whereas the experienced practitioner knows that, at best, they are emerging disciplines.)

An example might be long-term planning. A careful look at the practice in the United States shows that the assumption that long-range planning is a widespread practice here is false; although many talk about long-term (20 years and more) planning, few organizations actually practice it in any but the most simplistic forms. In short, we do not systematically practice what we preach. In thinking about peddling long-term planning to developing countries (where it may be needed even more than in developed economies), this gap between our rhetoric and the reality of our practices must be accepted and understood.

Those responsible for transfer must be aware of such problems, inconsistencies, and controversies about the technology itself before embarking on its transfer to a new and different environment. (This is a humbling experience, which I believe is all to the good as it helps the expert lower expectations for the transfer and to pay more attention to the factors that will induce error or unexpected results.)

Culture

The second group of obstacles in transferring a soft technology overseas is the cultural differences between the originating and receiving environments. "Culture" is a word with many and varied meanings, of course, but its use is limited here to those norms or characteristics of a working society that set it apart from other societies. These range from religious practices to acceptable social behavior to concepts of the appropriate role of the individual, groups, or the state.

The foreigner responsible for technology transfer, whether hard or soft, should be as knowledgeable as possible about the culture into which the technology is expected to fit. It is equally important, however, that those responsible for the transfer know as much as possible about the cultural quirks of the society (usually their own) from which the technology is taken.

Achieving such insight is a very different challenge, for the ability to see ourselves clearly is difficult if not impossible. (Robert Burns's poetic conclusion that it is a gift much sought after but seldom granted is exactly right in my view.) Whereas few truly understand their own society's peculiarities, we are all far more affected in our working life by our cultural makeup than we usually realize. Without having a cultural mirror forced upon us, we unconsciously assume that our quirks are a universal norm, or at least preferable to others, and are therefore not aware of their impact. Doesn't nearly everyone work a 40-hour week, usually 9 to 5 on Mondays through Fridays? Isn't planning before acting always preferable? Aren't profits the primary goal of any business or industry?

The answer to all of these questions in much of the developing world is no. Accordingly, any technology taking these norms for granted will not be easily transferred, for all technologies (and especially soft technologies) are rooted in the particular environment that spawned them. Logic suggests that technologies should be "deculturalized" before transfer, but this is seldom done or even considered. Careful analysis would occasionally reveal, however, that the technology is so rooted in one culture that it cannot be transferred at all, or at least not without major alterations.

These factors are often overlooked because we do not think that concepts such as strategic planning are a product of culture, but rather are simply a matter of common sense; the question is not *whether* to do it, but how. Transferring a technology such as strategic planning to an environment where those who will use it may implicitly believe that the future is foreordained, to one extent or another, assures that the transfer will fail or be greatly weakened. (What conceivable good is planning when the future is determined by God, not people?)

Having done all one can to learn about the new environment, however, the next responsibility of foreigners trying to transfer technology is simply to relax and be themselves. The best advice I ever received on this subject is "Be an authentic American; don't try to be a native."

Rather than trying to meld completely into another culture, the foreigner is better advised to learn what he or she can, starting with the social norms and courtesies, and how best to convey the universal, fundamental values: decency, warmth, family, integrity. Beyond that, the wisest course is also the easiest: be an open and honest representative of your own culture. Everyone knows about Americans, or Europeans, or Orientals (or at least they think they do), so making your obvious heritage harder to read by mixing it with superficial behaviors from the local culture easily generates confusion.

In our eagerness to learn and adapt, it is natural to try to act as much as possible like the citizens of the new environment ("When in Rome, . . ."). Complete success, however, is seldom possible, We have all known people who immigrated many years ago to our own society, as pluralistic as it is, but who still act in patterns we recognize as foreign and who are not aware of some of the cues that we instantly recognize. Only their children become completely "fluent" in the society.

An approach I have used is to try to implicitly create a "third," or neutral, culture in dealings with developing country colleagues. One can often slip into this, over time, with the knowledge and tacit cooperation of such colleagues. Courtesy and not seeking the dominant role are virtues that move across cultures very well where the working relationship is cordial and cooperative.

Practicing mutual deference and softening the more difficult edges of one's own cultural behaviors, or language—and with your counterpart doing the same—can create a combination of the two cultures that amounts to a third, or melded, culture. This meld helps keep your dealings on a level plane, reducing the likelihood that either side has an undue advantage in terms of values or cultural sophistication.

Other cultural barriers are rooted in behaviors or beliefs. Local practices may make what is easy in one society difficult in another. Disaster awaits those who would tamper with the workday in a country where, for good reason, the main meal is taken at midday, sensibly followed by a nap.

> In South Asia in the 1960s, U.S. AID and the Peace Corps tried to introduce row planting, perhaps the easiest and most nominally inexpensive "new" agricultural technology. Local farmers were carefully told what was required and of the increased yields and other benefits that would ensue.
>
> Planting in many areas there, however, was done by the women, and they were often overlooked—or perhaps lacked enthusiasm for a cost/benefit equation that required them to give up the socialization of working in clusters as they talked, a practice since time immemorial.

Communications

Because soft technologies are, by their very nature, conceptual rather than tangible, their transfer is uniquely dependent upon the ability to communicate what is involved. There is usually no model or demonstrator to illustrate how the technology works and exactly what the result is. Whereas it is relatively easy to grasp what a grain thresher or computer can do and how that might apply in a new environment, soft technologies require clear, unambiguous communication and, to a large extent, a leap of faith as well.

Successful communications depend on the ability of the transferor to articulate the concept, methodology, or procedure, and its benefits clearly, followed by near-perfect understanding of that communication by the recipient. This is not easy, of course; for proof we need look no further than the misunderstandings we regularly experience without the added burdens of distance, language, culture, and greatly differing circumstances.

The accepted wisdom is that language facility is the most powerful tool available to those working abroad in any capacity involving technology transfer. I certainly agree, but the language demands on those dealing with a soft technology can be more than, say, what is needed by two chemists discussing a process technology. Language deficiency is a significant handicap, but it is *not,* in my experience, the primary cause of communications failures when working abroad. The more fundamental communications flaw is a failure to assure that what is understood by the technology recipients is exactly what was meant by the technology transferor (and vice versa).

Talking about complex and often subtle concepts, as opposed to demonstrable hardware, requires virtually total fluency in order to be able to work as well in the foreign language as in one's own. (The Foreign Service Institute defines full fluency as the ability to communicate as well as a university educated native speaker. Years of study and immersion in the language are needed to achieve that level, of course.) Where that is not the case among all of the parties involved (the usual situation), then meticulous care is needed to ensure that communications are clear and well understood. Fortunately there is a great deal written on the subject of language as a barrier for working in technology transfer to a foreign country.

Slowing down communications, forcing an uncluttered attention to the basics, and paying close attention to detail are the best insurance that problems are not being overlooked. The foreigner is often better off remaining the outsider, conscious of his or her limitations (including language deficiencies), than falling into the belief that language fluency will solve communications problems.

At the risk of alienating any linguist friends, I contend that a language handicap should not intimidate those trying to transfer soft technology. Demonstrated competence, commitment, and a determination to overcome the problems in technology transfer can and do succeed. I speculate that a person comfortable and at ease in alien cultures, but totally bereft of language skills, would have a more enjoyable and productive time, for example, in making his or her way from Leningrad to Vladivostok than a trained Russian speaker who has never been abroad.

> In Nepal we had a Kansas farmer on our staff who was far more effective in promoting agricultural innovations with the local farmers than any Nepalese-speaking agronomist we might have had. The reasons were his enthusiasm, credibility, and conviction, all born of first-hand experience; qualities that he easily and abundantly conveyed with only rudimentary language skills.

Because nothing short of fluency is completely adequate, and few, if any, of the experts in a particular technology being transferred will be fluent, the choices in overcoming the language obstacles are limited. Often circumstances dictate the need simply to muddle through with whatever language capabilities one has, coupled with a recipient who is relatively fluent in one's own language. Although clumsy, this has the advantage of leveling the playing field and forcing both sides to concentrate very carefully.

Where language is a clear obstacle, then a translator is, of course, necessary to assure clear communications. In the best case, there would be people fluent in both languages on each side—among the foreign experts and also among the recipients—to avoid putting an undue linguistic burden on one person. Translators can become a barrier themselves, of course. In some cases, they illustrate the maxim that a little knowledge is a dangerous thing; the translator becomes involved in explaining or commenting on the technology itself rather than simply performing the narrow but critically important role of seeing that both sides understand each other.

Whether or not one uses translators, a minimal language facility in the courtesies of the other environment is fairly easy to acquire and makes the foreigner's acceptance easier. Not least, it demonstrates that the foreigner cares enough to be polite, even if he or she is unable to conduct professional discussions in the recipient's language. (Although not related to a technological purpose, John Kennedy generated wild enthusiasm in Berlin with a command of only four words of German, *"Ich bin ein Berliner."*) The foreign adviser should take the time and effort to acquire these social basics in the language, at a minimum, and polish them through use to become as graceful as possible.

For those who find foreign languages very difficult, plunge in with whatever ability you have—hurling grammar, tense, syntax (and your pride, if need be) to the winds. The very real linguistic advantage of any foreigner is the fluency of the native listener. Once used to the perhaps bizarre speech patterns of the visitor, the native's innate skills can sort out the meanings from context—if he or she is interested, committed, and sympathetic to the foreigner's efforts.

> It has often been observed that (New Yorkers and Parisians excepted) most people will go to great lengths to understand and assist the foreigner who is trying to use their language—out of appreciation for the fact that they make the effort, even at the risk of appearing the fool.

In many Third World countries, English is a strong second language. But even in those fortunate circumstances (for those of us who are native English-speakers, at least), misunderstandings are the norm, not the exception. Whereas most of the reasons for miscommunication are well documented in the literature on communication, negotiation, and the educational process, two additional factors stand out in the international arena:

- The foreigner's lack of fluency in nuances and local styles. (Everyone has had the experience of thinking we know what's going on, and later discovering that we were really getting only the main drift; the more removed we are from our own cultural milieu, the more likely this will be the case.)

- The unwillingness of the technology recipient to closely question the "expert." (This universal reluctance to question authority is aggravated by the Third Worlder's tendency to defer to the supposed expertise of the more cosmopolitan West.)

So far I have assumed that the expert responsible for the technology transfer is on-site in the receiving country. Although this is very desirable, it is not always the case; some transfers are attempted at long distance. Even those where the initial transfer has taken place with the experts' help in person, follow-up help is needed after the experts have gone home. This means that communication must be by mail, telex, fax, or telephone.

These communications have to be very carefully structured to avoid misunderstandings. Strict disciplines to promote clarity and conciseness, and to avoid ambiguities and sloppiness, are well worth the effort. To the extent possible, communication should be in writing, with the primary burden being on those who are using their native language. (This allows the other side to limit its communicating to yes and no kinds of responses.) Writing is preferable in order to be able to edit for clarity; to allow the recipient to be able to study the communication, not just react to it; and to have a record for reference when confusion or misunderstandings arise.

Behavior

Because the transfer of conceptual technologies is people-centered, the behavior of those involved—at minimum the individual "expert" and the primary recipient—is critical to the success of the transfer and its application. (A number of people are usually directly involved in any technology transfer, of course, but for the sake of simplicity I refer to the sides of the transfer as though there were only two.)

In any difficult transaction between foreigners, there is a tendency to behave unnaturally; that is, the parties act differently than they would in a familiar, strictly domestic situation. The foreign expert in the technology becomes aware (if he or she was not before) that what to them is a pretty simple matter is much more difficult to convey than in the "home" environment. In an unfamiliar place it becomes easy to see the people they are dealing with as difficult.

At that point all of the other obstacles can come into play: assumptions, language, and culture ("why don't they understand?"). This can quickly lead to blaming the local environment, or the citizenry, for the failure to communicate adequately.

The recipient is not much more comfortable. He or she may be intimidated by the technology, the expert, or both. That can lead to overly deferential behavior, or arrogance born of insecurity, or anything in between. At the least, when it is apparent that the new technology will require changes from present practice, the transaction begins to look like a problem rather than a solution.

Another factor affecting this dynamic in developing countries is that the recipient most closely involved with the foreign expert is often someone educated in the developed countries. Whereas most speak favorably about their years abroad, few such residencies come without wounds, particularly among young adults. Encountering hostility as a foreigner, struggling with a foreign language, social slights, and even racist incidents are common experiences for the foreign student abroad. It is therefore all too frequent that they, consciously or not, "give back" some of that experience to foreigners, in particular the visiting "expert."

None of this is new, however, or limited to technology transfer. All of us have experienced similar reactions in unfamiliar surroundings, or when dealing with an authority who has either knowledge or power that we do not. The primary point is that the atmosphere in which most soft technology transfers are expected to take place can be, and often is, marked by people acting under stress rather than behaving naturally.

I have noted that the stress of the transfer situation can result in defensive or apprehensive behaviors. It does little good to say "relax, you're only cutting your own throat," so what can one do? Some approaches have been mentioned: not trying to be something you're not; working to level the playing field, rather than allowing one side to dominate the transaction; and being sure of the technology's requirements, as well as its weaknesses and limitations.

Most important, given the amount of work and patience required to transfer soft technology successfully, the working situation should be as informal and cooperative as possible. The foreigner responsible for introducing the new technology should have confidence in his or her knowledge of it, what it can be expected to do, and what is necessary to achieve effective implementation. That confidence will go a long way to allowing the relaxation and concentration that is required for an effective transfer.

The stereotypic view of a transferor's nationality in the mind of the recipient may also be very relevant. At the risk of a vast overgeneralization, I suggest that North Americans have, by and large, a benevolent reputation in much of the developing world, whether deserved or not. They are typically seen, it would seem, as rather bumbling but well-meaning oafs. The primary reason may lie in how they are *not* seen; as colonialists (except possibly in Latin America and the Philippines) or exploiters, or as having overtly political designs. Whatever the truth, this general reputation for open, informal dealings gives North Americans an advantage when working abroad.

Lacking direct knowledge, I cannot speak with as much conviction about Europeans or Orientals. In general, they have an enviable record in working successfully in the developing world. This is probably because of their often greater knowledge of the developing country environment, their dedication to acquiring language skills, and perhaps their better awareness of the emerging interdependent world of technology, resources, and commerce.

No soft technology can work without the best efforts of both the provider and the recipient. Taking a "stand back and I'll show you how to do this" approach only proves that the expert can make the technology work; it does not instill much confidence in the recipients and can be counterproductive. Involvement in the bridging mechanisms between the expert and the user is a necessary element to any successful technology adaptation. This has recently been illustrated in our own society with the proliferation of computer technologies.

With the rapid increase in computer applications in the workplace, first in the more developed countries but increasingly in developing countries as well, we have discovered the need for a continuous and effective dialogue between users and experts. The users were, at first, technicians, clerks, and typists; now they include managers, planners, designers, and schoolchildren. The experts are the hardware and software designers, programmers, and manufacturers.

A decade ago it became apparent that the computer's capabilities were outstripping the ability of the users to conceptualize or effectively harness the power of computer applications to the business, government, and institutional worlds. The result was frustration, with the full capabilities of computer technology being largely underutilized for some years. (The easier, numbers-crunching applications had been less difficult, of course.)

Now we are beginning (but only beginning) to see the promise of computer-assisted *soft* technology applications. What is happening now is a relinking of the marketplace with the resource; the dialogue between the users and the experts is becoming increasingly effective. The users now realize that *they* must decide how the new computer technologies might help them in their work. They cannot (and need not) grasp the experts' potential abilities to meet those needs. However, in terms of applications, only they can see the needs. In turn, the experts now realize that they cannot fully understand potential applications and that their role is simply to adapt the technologies to fit the uses as those uses are made clear to them.

For each to be effective, a close and continuing dialogue between users and experts is vital. This is partly because each side is still a bit crippled; users will tend to think narrowly about how to automate existing practices, not how to design new ones, and computer experts tend to think narrowly about technical capabilities rather than about human-based systems as a whole.

In short, the evolution of computer applications becomes a cooperative effort, each participant acting in accordance with his or her own interests and abilities. This seems exactly the case for successful soft technology transfer. It cannot adequately be done from one perspective alone; each party to the transaction must take fully responsibility for his or her own role and function.

Apart from these specific behavioral observations, two overriding human qualities are required on both sides of the transfer equation:

- Patience and humility about the gap between theory and reality, and in making a successful adaptation to a specific new environment.
- A sense of humor and balance about the attempt to apply quasi-scientific disciplines—that is, these soft technologies—to real-world human and institutional affairs.

THE TRANSFER PROCESS

There is nothing remarkable about the transfer process for soft technologies, compared to the transfer of hard technologies, except that it is a slower process and calls for meticulous attention to detail. The first consideration in trying to assure the successful transfer of any technology, hard or soft, is looking beyond the idea of simply selling the technology itself. Both the technology provider and the recipient must be committed to full and successful implementation of the technology, not just nominal use. Although this seems obvious, the painstaking steps required for full implementation are often overlooked. (The reason for this shortsightedness is probably the fact that organizations, and the marketplace, typically reward "sales," not utilization.)

A major U.S. electrical equipment manufacturer had its reputation suffer in Saudi Arabia some years ago after selling the Saudis several generating turbines built to run on multiple fuels, including crude oil. The reason was not technical; the turbines would run on crude, as advertised, but only with careful maintenance.

Too late the company realized that it should have refused to sell just the equipment without also training and supervising the maintenance personnel so that the turbines (and the company's) reputation would not be jeopardized by the maintenance failures that eventually resulted in the ruin of some of the turbines.

The transfer procedure itself is similar to any project activity: careful design of the process, learning about each of the elements involved (the technology, its needs, intended uses, and the receiving environment), developing the resident support systems, midwifing the transfer itself, and follow-up. Expanding all project time-lines and personnel budgets by some multiple of what is theoretically feasible in a domestic transfer is also advisable.

Another important precaution is to have fallback alternatives for each element of the transfer project in order to be able to shift gears quickly when things go wrong. Murphy's Law lives and thrives, particularly in the Third World.

The transfer of soft technologies should not be left to experts or only one side of the transaction. It must be a collaborative effort between those who know the technology and those who are going to use it. Building this partnership is fundamental but difficult if the provider is overconfident about the technology or the recipient hs unrealistic expectations regarding the results. The expert is, after all, only an expert on the technology itself, not on its use or its usefulness.

It may be futile, and certainly is more than a little presumptuous, for the technology provider to take sole responsibility for adapting the technology to the local conditions. To bridge the gap between the two requires collaboration; the provider cannot presume to bridge the gap alone, but the provider and the user can. (In short, it is not the provider's "bridge," nor the user's "gap.")

I refer often to the technology "expert." In hard technology transfers, a technical expert typically is involved, for the hardware part of the transfer at least. Soft technologies, however, including the skills that need to accompany most hard technology transfers, are less technical and there often is no "expert" as such. For most soft technology transfers, the provider is often simply an experienced *user* of the technology. This provides an advantage, if capitalized upon, in that the recipient and the provider may have similar mindsets and regard the technology from the same perspective.

Having painted a somewhat gloomy picture of the problems associated with soft technology transfer, here is the good news:

- Successful transfers do happen, it is just not as easy as it sometimes looks. Not surprisingly, many of the successes occur gradually, by a process rather like osmosis, without the direct help of experts. In fact this is a preferable course where the transfer is not urgent; allow the technology to simply grow over time, rather than being force-fed by design.
- A fully successful transfer is not always needed. Particularly in developing countries, a 100 percent solution is often not necessary; only an improvement on the existing situation, which justifies the investment.

(This is an example of the relativity of the development process; moving directly from the oxcart to the auto is significantly different from moving first to a horse and buggy.)

- A lot of technologies can be adopted in parts, rather than absorbed all at once. This is what some literature calls "de-tuning," or disaggregation, that is, the fact that pieces may work where the whole would not. This is often preferable in any environment but should be considered very seriously where wholesale transfer promises to be difficult.
- Barriers to international transfer have been detailed, but an important positive factor is that there may also be an absence of some of the problems usually associated with innovation; obstacles that exist in the originating environment may not be present in the receiving environment or may be much less troublesome.

Because the process for international technology transfers is primarily driven by the differences in the originating and receiving environments, those responsible for transfers must work to widen their vision and sharpen their hearing in order to increase the likelihood of successful adaptation.

Any project having a significant level of uncertainties (which is nearly always the case with soft technology transfers) calls for sound planning and the use of more feedback and correction loops during the course of the project. Most failures in technology transfers are caused by issues that were unanticipated, so information becomes critical. For this reason, the technology provider wants to seek out questions, critical comments, and doubts about what is being proposed, for whether or not the doubters are right, their critical perspective can help locate potential sources of error. Accordingly, the key element in soft technology transfer is designing a transfer process that anticipates obstacles and successfully adapts the technology to match the receiving environment.

Chapter 9

International Transfer of Organizational Technology

D. Eleanor Westney

INTRODUCTION

The transfer of the social technologies of organizational structures and processes across societies is assuming growing importance in today's world. The literature on the international transfer of physical technologies has long recognized the importance of the accompanying transfer of organizational routines and skills. Recently, however, the transfer of purely organizational technologies—that is, of organizational structures and processes that do not involve physical technologies, or involve them only peripherally—has become an independent focus of concern for managers and management researchers alike. Managers are most concerned with *how* to transfer organizational technologies (such as quality control systems or personnel development programs) from one country to another. Researchers are becoming increasingly interested in developing conceptual frameworks for assessing the limitations on such transfers and with analyzing the processes of transfer and their effects.

One reason for the growing interest in the subject is the increasing internationalization of service industries, especially financial services, in which such organizational technologies as personnel development programs, task specialization, and structures for communicating and adding value to information constitute the core technologies of the firm. Another is the increasing salience of multinational corporations who consider that key elements of their competitive advantage lie in the way they organize tasks and people, and who therefore make extraordinary efforts to transfer a considerable array of these organizational technologies to their overseas subsidiaries. And there are multinational enterprises that fit neither of these categories but are finding that the growing integration of their operations across borders has put increased pressure on them to standardize at least some of their organizational structures and processes internationally—an effort that means transferring organizational technologies abroad.

The international transfer of organizational technologies is, of course, not limited to transfers from one part of a multinational corporation to another. With the heightened global competition of recent years, there is growing interest in the possibilities of emulating the organizational technologies of successful foreign competitors, even when these technologies have been developed in a very different society. And in the larger policymaking arena of the nation-state, where the concept of comparative advantage has been extended to include institutional factors such as educational and legal systems and industrial policy, there has been renewed interest in "learning" from the organizational structures of other societies, which are perceived as more successful in dealing with the new era of global competition.

Transferring social technologies across borders is even more complex than transferring physical technologies. Because organizational technologies are less codified than physical technologies, and because they are even more interdependent with their social context, their transfer from one society to another inevitably entails change and innovation, planned or unplanned. The study of the transfer of organizational technologies can therefore highlight issues in technology transfer that the focus on physical technologies has tended to neglect. And looking at that subset or organizational change and innovation that can be defined as the international transfer of organizational technologies can contribute a new perspective on the field of organization theory. This study draws extensively on the historical experience of Japan, a society that has been noted for its experience in transferring and adapting foreign organizational technologies, to address these issues.

TRANSFERRING PHYSICAL AND ORGANIZATIONAL TECHNOLOGIES

Max Weber likened bureaucracy, with its formal rules of procedure and specification of tasks, its codified patterns of communication and record-keeping, and its hierarchy of offices, to a machine. And like machine technologies, formal organizational structures can be transferred across societies. The spread of organizational technologies across societies is analogues to two related bodies of literature on international flows of technology. One is the literature on the diffusion of technology, which is analogous to the adoption and adaptation of foreign organizational technologies based on general models (that is, generalized concepts of a specialized organization performing a particular set of tasks, such as the factory or the newspaper), where there is no active involvement of a foreign organization that serves as a model. The second is the literature on technology transfer, when a specific organizational model is selected or imposed and there is active involvement by both the model (the "originator" in the language of the technology transfer literature) and the emulating organization (the "adopter" or "recipient"). It is, of course, the latter on which this chapter is building and to which it aims to contribute.

Some of the literature on international technology transfer recognizes the varying importance of the organizational aspects of the transfer. A recent review of the field, for example, cites Hall and Johnson's distinctions among product-embodied, process-embodied, and person-embodied technologies, and puts forward the proposition that the latter two are more difficult to transfer across nations than product-embodied technologies "because cultural differences at the organizational and societal levels play a greater role in such transfers."

In such approaches, however, the focus is still firmly on the physical technology that is "embodied" in the social. An alternative approach is to extend the field to include purely organizational technologies such as new organizational forms and ways of organizing people and resources. The field would then cover four broad categories:

1. Physical technologies (akin to "product-embodied technologies," such as FAX machines).
2. Physical technologies that are supported by certain organizational technologies (such as telephone systems).
3. Organizational technologies that are supported by certain physical technologies (such as postal systems).
4. Purely organizational technologies (such as T-groups or quality control circles).

The latter two categories are included in the following discussions of organizational technologies.

The international technology transfer literature has focused its analysis on three major sets of issues: the selection of technologies to be transferred; the modes of transfer, and the factors that lead to success or failure in the transfer (with all the difficult conceptual problems involved in defining success or failure). In each area, the analysis of organizational technologies both benefits from the insights of the work to date and demands the consideration of an additional range of variables.

Selection of Technology

The discussion of the selection of technologies for transfer across societies has generally been dominated by economic models that emphasize rational choice and the calculation of costs and benefits.[1] These have been increasingly supplemented by a focus on the strategic competitive context and by some attention to social context, especially access to information.[2] The analysis of the selection processes for organizational technologies demands much greater attention to social context and to the social rather than the economic aspects of strategic content. The greater uncertainty in linking specific organizational structures and processes to economic outcomes and the lack of precise conventions for measuring the economic efficiency of organizational makes the greater importance of social factors inevitable.

We should note that this is not necessarily the case for organizational technologies that support specific physical technologies, where the transfers are often made without conscious selection. They are simply defined by both originator and adopter as the way things are done. A recent discussion of the consequences for Australia of its long-standing dependence on imported technology states this vividly:

> "What have been imported, therefore, are not only the technological systems, but also the organizational assumptions and designs that are implied by those systems—a social and cultural milieu that remains unpacked when the technological baggage is delivered."[3]

In cases where the organizational technologies are themselves the focus of the transfer, however, the issue of how the technology is selected becomes as relevant as it is for physical technologies, although three factors assume much greater salience: access to information, the prestige of the originator, and the impact on the power and position of key individuals and subgroups in the adopting organization.

Access to information has also been recognized as a significant factor in the international transfer of physical technologies.[4] It is even more important for organizational technologies, in part because they are less likely to be actively hawked in international markets. Within multinational corporations, alternatives to the technology owned by the parent are not often considered in transfers to subsidiaries, whether the technology is physical or organizational. Alternative external sources of technology are somewhat more likely to be considered for physical than for organizational technologies, however, largely because detailed information on alternatives is easier to obtain. Where transfers are across organizational as well as national boundaries, access to information is likely to be strongly skewed by the direct experiences of key individuals in the adopting organization, such as foreign study, diplomatic contacts, or postings abroad. The lingering legacy of colonialism as a factor in transfers of organizational technology is closely linked to access to information.

The prestige of the originator is a factor that may be at odds with access to information: the former is structured and relatively slow to change, the latter is often serendipitous. However, given a choice, the adapter is more likely to seek out information from prestigious organizations. DiMaggio and Powell have pointed out that where uncertainty over the effectiveness of alternative organizational forms is high, organizations are likely to adopt the patterns of other organizations that have the reputation in their immediate environments of being successful.[5]

Finally, the impact of imported organizational technologies on the power and position of key individuals and subgroups within the adopting organization is obviously important, given the difficulties of demonstrating conclusively that a given organizational technology is unquestionably more effective and efficient than alternatives. The impact is shaped not only by the direct consequences of the technology itself, however, but also by the potential halo effect on the reputation of adopters. For this reason, even those whose power may be adversely affected by a given foreign organizational technology may become its advocates if they believe that over the long run the reputational advantages of being identified with the change will be high. The "political factor" may therefore also be related to the prestige of the originator.

It is worth repeating that these three factors are not unique to the transfer of organizational technologies; they are also involved in physical transfers. As analyses of organizational transfers clarify these factors and the relationships among them, and undoubtedly add additional social factors to the list, we may begin to amplify our paradigms of the transfer of physical technologies as well.

Modes of Transfer

The discussions of the modes of international transfer of physical technologies would seem at first to have little relevance to organizational transfers, largely because they have been centered to such an extent on the ownership of technology and the fair allocation of the returns from transferred technology, both of which are virtually impossible to establish for organizational technologies. But whereas the detailed discussions of modes of transfer (from licensing to setting up a wholly owned subsidiary) do not map directly onto an analysis of the transfer of organizational technologies, the factor underlying the discussion—the relationship between the provider of the technology and the adopter—is decidedly relevant.

In organizational technology transfers, there is enormous scope for variation on several variables: the locus of initiative for the transfer, the extent of the provider's involvement in the transfer, and the extent to which the originator is affected by the outcome of the transfer process. In many transfers, the initiative comes from the recipient organization, which takes virtually all the responsibility for the transfer; the provider may not only be unaffected by the outcome, but (except for a few key individuals) largely unaware that a transfer is occurring. At the other extreme—for example, within a multinational corporation, or an imperialist regime that is imposing the patterns of home country administration on its colonies—the adopter may be a most unwilling partner to the transfer. The transfer may involve a few hours of the time of some key informants in the providing organization, as they host a study mission and provide documentation, or it may involve multilevel exchanges of personnel between the two organizations.

A second aspect of the modes of transfer that has been noted in the literature on physical transfers but is much more salient in organizational transfers is whether the technology is transferred as a complete entity, a "package," or whether it is transferred only in elements, or "fragmented form."[6] One of the most salient differences between organizational technologies and physical technologies is the extent to which elements of the technology can be uncoupled and adapted independently of the whole. In other words, there are two broad categories of the transfer of organizational technologies: the transfer of entire organizational forms and the transfer of elements of an organization.

The transfer of entire organizational forms is most widespread during periods of major social and political transformation, such as the widespread emulation of Western organizational systems during the Meiji period (1868–1912) in Japan, and the adoption of Soviet models of economic, political, and military organization during the early years of the People's Republic of China. On a far less dramatic scale, however, such transfers occur fairly frequently in today's world: the establishment of a new embassy, a branch office of a trading company, or a new McDonald's outlet overseas all involve the transfer of a complete organizational form. Transfers of complete forms occur primarily when new organizations or organizational subunits are being established, although this is not always the case. In some instances, for example, acquisition of a local company by a foreign firm, the parent company may choose to impose its own organizational patterns on its subsidiary.

The second broad category is the transfer of elements of an organizational form, usually—although not always—into an existing organization that maintains its overall structure but tries to incorporate a new subsystem or element. The transfer of production management systems from Honda to its British strategic ally, the Rover group, is one example of a growing number of cases where Japanese organizations are transferring quality control and production management systems to allied organizations, subsidiaries, or (as in the case of Fuji-Xerox) parent companies abroad.

Obviously creating these two categories—the transfer of entire organizational forms and the transfer of elements of organization—is somewhat more arbitrary than is the case for physical technologies. The complete organizational form of the original model is never perfectly replicated in another society, however earnestly the architects of the transfer might desire such an outcome; some elements of the organization's structures and processes will of necessity derive form the local environment (a more detailed discussion of this process follows). And in many cases where the transfer of organizational technologies is expected to focus on a very limited subset of the structures and processes of the original model, the actual transfer extends well beyond what was first planned. The distinction in practice is between the case where the organizational form is transferred and elements of the organization are changed in response to the demands of the local social context, and the case where elements of organizational structure and process are adopted into an organizational form that differs from that in which the elements originated. This distinction captures one of the key organizational dilemmas of the multinational corporation: whether to try to transfer the organizational form of the parent organization to its foreign subsidiaries, or whether to adopt local organizational forms and transfer only those elements of organization needed for smooth coordination within the system.[7]

Success and Failure Factors Vs. Innovation Processes

The analysis of success and failure and of key success or failure factors is much more problematic for organizational than for physical technologies. Indeed, the problems suggest that a more fruitful way to analyze outcomes is in terms that are much more salient in the study of international transfers of organizational technology than in transfers of physical technologies: departures from the original technology, either at the time of transfer or shortly thereafter, in response to the different social context into which the technologies are being introduced.

In the literature on organizational aspects of international transfers of physical technologies, the most powerful influences in modifying foreign-based organizational patterns are ascribed to national culture, usually couched in terms of the social-psychological predispositions of individuals. However, a somewhat more precise characterization of the cultural influences on organizational transfers can be drawn from the set of international transfers that occurred during one of history's most dramatic periods of cross-border organizational technology transfer, the establishment of new organizations on Western models in the 1870s and 1880s in Japan.

THE CASE OF MEIJI JAPAN

In the 1870s, Japan embarked on a wide-ranging transfer of organizational technologies from the industrialized nations of the West. The French army, public education system, police and legal system, the British navy, postal and telegraph systems, the U.S. banking system and the American Chamber of Commerce—an extraordinary range of foreign organizational forms— were transferred to Japan. In the late 1870s and the 1880s came a wave of reorganizations that in many cases involved a switch to a new foreign model: the German army instead of the French, the Belgian centralized banking system in place of the American, first the American and then the German public school system instead of the French, and the German legal system.

The motivation for this massive transfer of organizational technologies was complex. In the case of the core state organizations, especially the military, where the first organizational transfers began even before the Meiji Restoration of 1868, the motivation was clearly to defend the country against the kind of foreign military incursions suffered by the Chinese and most of the rest of Asia. The inability of traditional Japanese military technology and organization to repel the Westerners had been demonstrated in some limited confrontations between the more militant domains and the British in 1862– 63. In a relatively short period of time, however, the young samurai who occupied key positions in the new Meiji government realized that a modern military system required an industrial and educational infrastructure that Japan lacked, and they turned to the West for the organizational models they needed.

In other areas, particularly the police, the judicial, and the legal systems, the adoption of Western organizational forms was a necessary condition for the abolition of the extraterritoriality clauses of the unequal treaties, which the Japanese found extremely irksome. And in still other areas, such as the postal system, ambitious young Japanese realized that creating a new organization using advanced Western organizational technologies was not only a way to contribute to building a modern nation that could win the respect of the advanced nations of the West, but also a means of rapidly advancing their own careers.

The mix of international and external pressures therefore varied across organizations. The wide scope of the adoption of Western organizational technologies legitimized the use of foreign organizational technologies, in part by defining them less as "foreign" than as "modern." It went so far as to generate a kind of internal pressure on existing organizational systems to look abroad for models.

The adoption of entire organizational forms dominated the transfers of the first decade of the period. Once the new organizations were established, however, subsequent efforts to improve them were roughly divided between the introduction of additional elements of foreign technologies (sometimes drawn from the original model, and sometimes from a different model in a different foreign country), and the replacement of the original model by an entirely different organizational form.

The extent of the involvement of the foreign organization in these transfers varied enormously. In some cases, including the navy and the army, there was a formal agreement between the foreign government and the Japanese government, providing for the dispatch of large military training missions to Japan and the reception of young Japanese into military training programs in the foreign country. In other cases, such as the railroads and the legal system, the Japanese government concluded agreements with private citizens experienced in the particular organizational technology to come to Japan as *oyatoi-gaijin* (literally, "employed foreigners"). In all, several thousand foreign advisers were employed during the Meiji period.[8] In still other cases, such as the postal system, the primary modes of transfer were focused study by a small number of Japanese abroad and the collection of formal regulations and published materials, with virtually no awareness on the part of the foreign organization that a transfer of organizational technology was taking place.

The international transfer of organizational technologies into Meiji Japan was on a scale unprecedented in modern history. It was strikingly deliberate and wide-ranging; it took place in a very compressed period of time; it drew on a range of foreign countries for models, not just on one; and it involved transfers that cross civilization as well as national boundaries. Not surprisingly, under these circumstances, imperfect information on the Western organizational technologies was a significant factor in departures from the original technologies. Japan in this era was particularly handicapped in gathering and processing information. The country enjoyed neither the contiguity of European societies, which allowed relatively easy exchanges of information, personnel, and organizational technologies across borders, nor the waves of immigration that carried information and skills across the oceans to the United States and the British dominions. In the 1870s, Japan faced not only geographical separation from the centers of organizational development that were the sources of organizational technologies, but also formidable language barriers. Of all the industrializing nations of the nineteenth century, Japan alone possessed a non-Indo-European language. Moreover, virtually no Japanese in the early Meiji period had extended personal experience of the models that the society's leaders were determined to adopt.

However, whereas Japan in the early Meiji period labored under particularly acute handicaps in garnering information on the organizational technologies it was transferring, imperfect information is in general an important factor in departures from organizational models in international technology transfers. Perfect information about an organizational form is never available to those engaged in using it as an organizational model. Selective perception on the art of the informants in the model organization as well as on the part of the information gatherers restricts the information available and renders inaccurate some of what is available. Particularly on the day-to-day functioning of the organization at the lower levels, where much of the organizational process is tacit rather than codified in formal rules, information may be missing altogether. The more numerous and multilevel the exchanges of personnel between the provider of the technology and the adopter, of course, the less significant imperfect information is likely to be.

When information about some aspects of the organizational technologies is missing, it is highly unlikely that the gaps will be filled in a manner that duplicates the features of the original. What will more likely fill them is an implicit model drawn from the past experiences of the new organization's recruits, many of whom (particularly those in supervisory and upper level positions) are drawn from other organizations in the immediate environment. They enter the new organization with a strong, but often not clearly articulated, model of what constitutes appropriate organizational structure. One of the major challenges in the international transfer of organizational technologies is therefore either to utilize this implicit model in its own structures or to replace it through aggressive training and socialization.

In Meiji Japan, many of the new organizations based on foreign organizational technologies tried to deal with this problem by formal training: the police and the postal system, for example, both set up training programs for their new recruits and for middle-level managers that considerably antedated any such education in the original foreign models. The explanation for this lies partly in the importance of the example set by the earliest foreign organizational technologies adopted, the army and the navy, which established a number of training institutes from academies for the education of very young officer candidates to courses for retraining senior officers.[9] In large part, however, it stemmed from the recognition of the need to replace older, more traditional models of roles and organizations with the roles and patterns of the new organizational technologies. In some cases, the training programs were phased out after a number of years of operation had provided a series of cohorts with the necessary training; in others, the programs became institutionalized and grew in scale and scope.

The changes in imported foreign technologies that result from imperfect information and strong implicit models are unplanned and often unrecognized, and therefore they are usually extremely difficult to trace. Unless they involve dramatic departures from the original model, neither the members of the new organization nor casual observers from the original model are likely to be aware of them. Departures from the original technologies that are deliberate and planned are much more salient. In Meiji organizations, these tended to be one of two types: (1) changes to preserve certain valued traditional patterns often had more to do with the organization's influence on the social environment than with the internal operations of the organization itself. For example, the Ministry of Justice objected to allowing dependent members of a household to open postal savings accounts without the permission of the head of the household,[10] and senior government officials in general insisted that those operating postal stations could not be females, as was often the case in the British model. Both of these examples involved maintaining the hierarchical household patterns that the Meiji leaders were anxious to codify in the new legal codes and inculcate through the educational system.

Concessions to the less-developed environment involved such features as modifying recruitment criteria and delaying the introduction of certain organizational functions (such as the parcel post). Departures of this type entailed an implicit assumption that the Japanese organization would move toward Western patterns over time.

Finally, a key factor in the Meiji adaptations of Western organizational technologies was a consequence of the fact that the institutional environment of the transferred organizational technologies differed significantly from that in which the original model was embedded. One useful way of conceptualizing the differences between the organizational environments is the model of the "organization set" put forward by William Evan.[11] He portrayed each organization as operating within a network of other organizations that provided it with needed inputs, received its outputs, and exerted regulatory control over some aspects of its operations (this last category includes, among others, state regulatory agencies, labor unions, and professional associations.) A manufacturing firm, for example, relies on other firms for its parts and materials, banks for its financing, railways, and transport companies to deliver its inputs and take away its products, a postal and telecommunications system to maintain regular contacts with customers and suppliers, and so on.

For most of the early Meiji organizational technologies transferred from the West, the original model was embedded in a set of organizations that often had no direct counterpart in their new setting in Japan. This situation was particularly marked in the case of Meiji Japan, but it is by no means unique to it. In the international transfer of organizational forms, one of the critical factors conducive to adjustments in the organizational technologies is the absence of one or more members of the original organization set. This may be because the adopting organization is located in a less-developed society, but it is not necessarily the case; the institutional landscape differs across the highly industrialized countries as well.

In Meiji Japan, the new organizations pursued one or more of four strategies:

1. Resort to local functional equivalents. The new organization turned to an organization or set of organizations that were already present in the new environment and could perform a similar function. For example, the early Japanese postal system, lacking the railways and the transport companies that supplied the infrastructure of the British postal system on which it was modeled, turned to the traditional transport guilds that had supplied relays of runner for the old post towns.[12]
2. Internalization. The new organization performed within its own boundaries the task that was carried out by a member of the organization set in the original. For example, given the relatively slow development of law courts in Meiji Japan, the police took over the tasks of imposing fines and prison sentences for a wide range of minor offenses that in its French model were handled by Judicial officials.
3. Elimination. The new organization modified its activities to do without the particular aspect supported by the missing part of the organization set. For example, early Japanese newspapers, in the absence of a population of business firms that were accustomed to advertising through that medium, relied primarily on sales revenue and politically motivated subsidies from wealthy patrons or political groups, rather than the mix of sales and advertising common in the Western press of the day.
4. Organization creation. The new organization acted as what Arthur Stinchcombe[13] has called an "organization-creating organization" and mobilized the resources to establish other organizations to perform the required activities. For example, in the early 1870s the Ministry of Public Works, finding that the shortage of trained engineers in Japan was a serious constraint on its ability to carry out the kind of projects that its European models were undertaking, set up the Imperial College of Engineering. Later the college was transferred to the jurisdiction of the Ministry of Education and became the Faculty of Engineering at Tokyo University.

The first three of these strategies (resort to local functioning equivalents, internalization, and elimination) all involve modifications of the imported organizational technologies. Adopting the first strategy means that the organization must create internal structures and processes for dealing with a different type of support activity and a different relationship with at least one part of its organization set. The second and third strategies entail a direct change in the activities performed within the organization, and therefore in its internal structures and processes. And the fourth strategy, organization-creation, often involves further transfers of organizational technology.

All of the preceding change factors—imperfect information, implicit models, deliberate changes to maintain valued patterns or to adjust to a less-developed environment, an incomplete organization set—apply equally to transfer of complete organizational forms and to transfers of elements of organization (although the preceding discussion, given the range of case studies in Meiji Japan, focuses primarily on complete organizations). But the transfer of organizational elements involves still another change factor. That is the intraorganizational analogue of the organization set: the organizational structures and processes that support the particular element in its original setting.

Each element of an organization is coupled (sometimes tightly, sometimes loosely) with other elements in the original organization, and when these elements are different from those in the adopting organization, they can exert a considerable pull on the imported technology, either for changes in the transferred element or for its rejection altogether. An example from the Meiji period is the introduction in 1879 of American-style elected school boards into the public education system. The goal of the education minister who initiated the introduction of this particular organizational technology was to increase enrollment in the schools by increasing community involvement through the elected boards. However, in the United States the local boards were part of a locally funded and directed school system, whereas in Japan the system had been modelled in 1872 on the highly centralized French system. As a result, the boards' functions were more circumscribed and the base for their influence considerably lower in their new setting. Moreover, their introduction met fierce resistance from Ministry of Education officials and from teachers, who were accustomed to a high level of autonomy from direct community influence. The elected boards were abolished in the following year.

CONCLUSION

These examples from Meiji Japan provide a beginning for creating a typology of change factors that shape the adaptation of transferred organizational technologies. The identification of such factors, and further research to assess their relative weight in different contexts, have important practical implications, helping those engaged in organizational transfers to anticipate more clearly the forces for modification of the imported technologies and thereby indicating how such adaptations can be planned for and accommodated. They can also play a role in advancing the fields of technology transfer and organizational theory.

Despite its longstanding concern with the effects of technology transfers on recipient societies, the international technology transfer literature has tended to focus more on the process of making the transfer deal and managing the initial transfer than with the analysis of the subsequent processes of modification and adaptation. The inclusion of organizational technology transfers more explicitly into the field will contribute to developing a clearer understanding of longer term adaptation processes and of the influence of social context on making and managing transfers.

Moreover, the potential contribution to organization and social change theories is also significant. Social scientists have long pointed out the critical importance of the "organizational revolution" of modernization, as more and more activities come to be performed in the context of formal organization rather than kinship or community groups. Several theorists have emphasized the importance of the sequence by which these organizations developed in later modernizing societies, as opposed to the first-comers. They have paid less attention to the extent to which the organizational technologies of the first-comers were transferred to later developing societies. In part this issue has been overlooked because both modernization theory and organization theory have tended to focus on variables within a society to explain the direction and speed of change. Adopting and adapting "rational" modern organizational systems was seen as a natural evolutionary process; what needed explanation, in this paradigm, was why this process was "blocked" in certain societies. Aspects of the development of modern organizational systems that were imitative of Western organizations were regarded as incidental to the more significant internal dynamics of what was taken to be natural social evolution. The inescapable observation that even organizations that tried to replicate closely the patterns of an advanced country model came to exhibit features that differed significantly from those of the original model has been widely taken as evidence of the primacy of intrasocietal variables (usually national culture) over external variables (in this case, foreign organizational technologies). However, a more precise analysis of the processes that shape the adaptation of organizational technologies to a social context that differs significantly from that in which they originated is essential to understanding not only the international transfer of social technologies but also the processes of organizational and social change.

Notes to Chapter 9

1. See, for example, Robert Stobaugh, and Louis T. Wells, Jr., *Technology Crossing Borders: The Choice, Transfer, and Management of International Technology Flows*, (Boston: Harvard Business School Press, 1984).

2. Michael Amsalem, "Technology Choice for Textiles and Paper Manufacture." In Stobaugh and Wells, *Technology Crossing Borders*.

3. Stephen Hill, "Technology and Organizational Culture: The Human Imperative in Integrating New Technology into Organization Design," *Technology in Society*, no. 10 (1988), pp. 234.

4. Amsalem, "Technology Choice."

5. Paul J. DiMaggio, and Walter W. Powell, "The Iron Cage Revisited: Institutional Isomorphism and Collective Rationality in Organizational Fields," *American Sociological Review*, 35, no. 2, (1983), 147–60.

6. See G. Vaitsos, *Technology Policy and Economic Development* (Ottawa: IRDC, 1976); and Abdelkader Djeflat, "The Management of Technology Transfer: Views and Experiences of Developing Countries," *International Journal of Technology Management*, 3, no. 1–2, (1988), 149–65.

7. Anant Neghandi, Golpira S. Eshghi, and Edith C. Yuen, "The Management Practices of Japanese Subsidiaries Overseas," *California Management Review*, XXVII, no. 4 (19XX), 93–105.

8. Hazel Jones, *Live Machines: Hired Foreigners and Meiji Japan* (Vancouver: University of British Columbia Press, 1980).

9. Ernst Presseisen, *Before Aggression: Europeans Train the Japanese Army* (Tucson: University of Arizona Press, 1965).

10. Yuseisho, *Yusei Hyakunen Shi* (One Hundred Years' History of the Postal System) (Tokyo: Yoshikawa Kobunkan, 1971), p. 158.

11. William Evan, "The Organization-Set." In James Thompson: *Approaches to Organizational Design* (Pittsburgh: University of Pittsburgh Press, 1966).

12. Kyozc Takechi, *Meiji Zenki Yusoshi no Kisoteki Kenkyu* (Basic Research into the Postal-Transport History of Early Meiji) (Tokyo: Yusankaku, 1978), pp. 15–17.

13. Arthur Stinchcombe, "Organizations and Social Structure." In James March, (ed.): *The Handbook of Organizations* (Chicago: Rand McNally, 1965).

Chapter 10

International Technology Communication in the Context of Corporate Strategic Decision-making

Richard D. Robinson

To be best understood, the commercial traffic in technology internationally should be considered in the general context of corporate strategic decision making. For what technology to communicate to foreign subsidiary, partner, associate, contractee, or consumer is a moot question. So likewise is the choice of the form in which the technology should move. By considering international technology communication in the more general framework of corporate decision making, one can relate it to both organizational theory and to the theory of direct foreign investment.

The process by which decisions are rendered in respect to the location of corporate functions—all of which contain technology—and to the extent to which technology is traded separately from capital and goods, consists essentially of five steps:

1. Identifying the separable links (that is, applied technologies) in the firm's value-added chain.
2. In the context of those links, determining the source of the firm's true competitive advantages, considering both economies of scale and of scope.
3. Ascertaining the level of transaction costs between links in the value-added chain, both internal and external, and selecting the lowest cost mode, which implies conscious steps to minimize those transaction costs, whether internal or external.
4. Determining the comparative advantages of countries (including the

firm's home country) relative to the production of each link in the value-added chain and to the relevant transaction costs.

5. Developing adequate flexibility in corporate decision making and organizational design so as to permit the firm to respond to changes in both its competitive advantages and in the comparative advantages of countries, both being highly dynamic.

Each of these steps requires further definition and expansion.

IDENTIFYING THE SEPARABLE LINKS IN A FIRM'S VALUE-ADDED CHAIN

A firm's product, whether it be "hard" (a good) or "soft" (a service including information and skill transfer), consists of a bundle of technology-driven activities—a value-added chain—some or all of which links (applied technologies) may be performed by the firm itself (that is, internalized). Others may be purchased from unrelated parties, or contracted out (externalized). It is often rewarding to look at each link that theoretically could be performed *geographically separate* from other links and by *separate firms*. The purpose of such examination is to determine wherein the real competitive edge of the firm lies. That is, which links are the real source of its profit? The more successful firms in the long run, one suspects, are not interested in maintaining control over a given function or set of functions simply for the sake of control (that is, power), but only if by so doing it is more profitable.

> A well-known food-processing firm, whose principal raw product source were company-owned plantations in less developed countries located within one region, eventually realized that its real competitive power lay in farm-related R&D, the transport of perishable agricultural produce, the marketing of such produce, and its ability to provide relatively low cost finance to producers. Actual production was something nearly anyone could undertake successfully, given technical direction, purchase contracts, and financing. Further, the foreign ownership of agricultural land was politically high risk. Best the firm externalize production and limit itself to the provision of relevant technology, distribution, marketing, and finance. The nature of the firm was thus redefined. That is, its boundaries were redrawn.

The value-added chain may be portrayed as in Figure 10.1, which is a much oversimplified representation in that each link consists of many possibly separable functions, only some of which are listed. Not shown are the

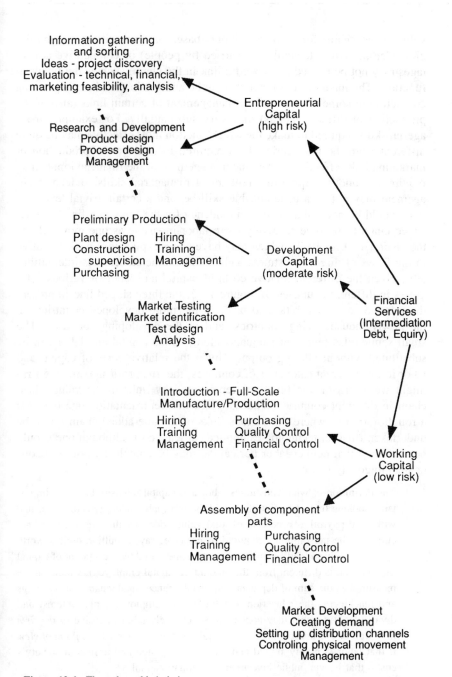

Information gathering
and sorting
Ideas - project discovery
Evaluation - technical, financial,
marketing feasibility, analysis

Entrepreneurial
Capital
(high risk)

Research and Development
Product design
Process design
Management

Preliminary Production

Plant design   Hiring
Construction   Training
   supervision   Management
Purchasing

Development
Capital
(moderate risk)

Market Testing
Market identification
Test design
Analysis

Financial
Services
(Intermediation
Debt, Equiry)

Introduction - Full-Scale
Manufacture/Production

Hiring   Purchasing
Training   Quality Control
Management   Financial Control

Working
Capital
(low risk)

Assembly of component
parts
Hiring   Purchasing
Training   Quality Control
Management   Financial Control

Market Development
Creating demand
Setting up distribution channels
Controling physical movement
Management

Figure 10.1. The value-added chain.

value-added chains for secondary inputs based on quite different technologies. Certain skills (technologies carried by people) or competitive advantages may not be related to a specific link in the chain, such as the financial function. The management function may be another in that there is growing conviction in some circles that the management of certain links (also of the production of different goods) may vary substantially. For example, managerial skills required to make the R&D function most effective may be quite different from the managerial skills required to manage the production or marketing links effectively. And the management of the financial inputs may require still another, quite different set of managerial skills. Indeed, management may not be a generalizable skill beyond a certain trivial level.

It should be noted that there is a tendency for the links in a firm's value-added chain to become relatively more labor-intensive as one moves down the chain (see Fig. 10.2), which may have some explanatory power relative to the flows of direct investment and of subcontracted services internationally, given the different relative costs of capital and labor of various categories in different countries. In Figure 10.2, the three sloped lines represent the relative costs of capital and labor in the more developed countries, in the newly industrializing countries, and in the developing countries. The small semicircles represent isoquants along which capital and labor can be substituted without altering output. Given the relative costs of capital and labor in the different categories of countries, the true profit-maximizing firm might well perform the functions represented by links in the value-added chain in different countries. That is, from this representation one may get a rough notion of where the various links of a value-added chain might be undertaken (that is, technologies applied) at least cost, although one should bear in mind that neither labor nor capital may be completely homogeneous or unambiguously distinct.

> The definition of what constitutes labor and capital intensity becomes important. Looking from the point of view of the firm, labor intensity is often equated with total payroll as a percent of total value added by the firm or, more accurately, the person-hours required to produce, say, a million dollars worth of output (value added). Capital intensity is measured by the amount of capital *used* (which is different from the amount of capital employed), which can be measured as the sum of depreciation, obsolescence, maintenance of buildings and machinery, the education and training of employees, plus interest and dividends. But, what may appear to be relatively labor-intensive *to the firm* may well be considered relatively capital-intensive *from society's point of view* if that labor is highly skilled and has been educated and trained at society's cost—that is, via public investment in human capital.

An element of confusion intrudes here, for the relative maturity of product and process must be considered separately at each stage of the value-added

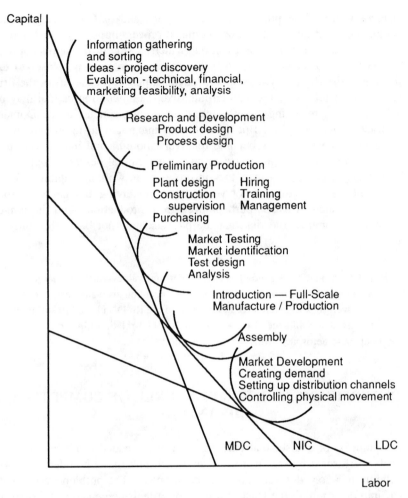

Figure 10.2. Changes in the value-added chain relative to comparative national advantage in respect to capital and labor.

chain. A product—such as a vehicle—may become mature and, hence, increasingly labor-intensive, but the introduction of an entirely new manufacturing process such as automation may render the process much more capital-intensive, for one must include the capital embodied in the human skills required to design, build, install, operate, monitor, and maintain the automated equipment. If the cost per vehicle built in an automated plant is less than the cost per vehicle coming out of the more traditional plant using the more mature (labor-intensive) technology, then the comparative advantage shifts back to the more developed country where the cost of capital relative to labor is low. The product is a mature one and may be changed only marginally, if at all, by the new automated processing.

By externalizing a link or function (which is simply the application of technologies) lower down on its value-added chain (that is, closer to the consumer, thereby implying a shorter payback period for the firm), the firm renders itself relatively less capital-intensive. Less working capital may be required before receiving the payback, for example, conducting R&D under contract *for another firm*. But if the firm is integrated all up and down its value-added chain, it may become relatively more capital intensive in that it must work much longer and commit greater capital before recouping its investment in payroll, in inventory in progress, in working capital. It must now wait many months, if not years, before receiving any payback from early links, such as R&D, perhaps even from production. And the longer the wait, the greater the risk that the payback may not be forthcoming at all.

> A firm producing a good of very high per-unit value—such as a large aircraft or computer—with many years from idea inception and sales (and collections) may well seek to externalize at least part of the R&D and production functions in a strategic alliance. Boeing Aircraft's initial relationship with a group of Japanese aerospace firms was of this nature.

DETERMINING THE FIRM'S SOURCE OF COMPETITIVE ADVANTAGES

The point is, of course, that each link in a value-added chain may or may not be performed by the firm itself, and each link may be undertaken in a different location, domestically or internationally. The problem resolves itself into determining the least cost mix, or at least perceived to be such by the firm's decision makers, both in respect to the activity embedded in each link and *the transaction cost involved*. This latter point is discussed further below.

Two concepts are useful in making the analysis relative to what to externalize and where to locate: (1) economies of scale and (2) economies of scope. The first is familiar to most. It has to do with the employment of functionally specific or product specific factors so as to achieve minimum unit production costs (that is, optimum intensity of use of such factors).

It is useful to think of these factors as representing fixed costs over a broad range of production and included within the production function in that they are identifiable with a particular activity or product and, hence, may be allocated unambiguously to that activity or product.

Economy of scope, in contrast, relates to the utilization of factors not identified with a specific activity or product, in such manner as to achieve the lowest overhead burden per unit of production and, thereby, gain an optimum spread (or scope) of use (Fig. 10.3). We refer here to those factors representing fixed costs over a broad range but not identifiable or attributable to specific function or product, and, hence, are not part of the production function. Though at the border, whether a particular economy results from scale or scope may be somewhat fuzzy, Figure 10.3 may be useful in separating the two concepts. It should be born in mind that reduction in cost resulting from "organizational learning" (i.e., moving down the learning or experience curve) can impact on both scale and scope, although *special practices developed by a firm to accelerate learning is an economy of scope, if* such practices can be applied effectively to *various* functions and/or products. And a firm's ability to transfer learning internationally with least cost may be an economy of scope of signal importance, likewise the learning of how to operate effectively within a specific country—which is a learned capacity. Both skills (technologies, if you will) are perhaps generally applicable. That is, they are not limited to a particular function and/or product and, hence represent a potential economy of scope.

A classic example of an economy of scope lies in the area of negotiating with a foreign government, which, given the importance of government intervention relative to the entrance of foreign business, is of utmost importance in international business. Negotiating skill may be a function of knowledge and experience, communication skills, personal attributes (i.e., employing the right people), training, and political leverage. Political leverage, in turn, may be a function of the corporate image, personal relations with key government officials, and the development of trust and confidence—all of which take time and practice to acquire. Such negotiating skill (a form of "soft" technology) may not be specific to a particular function or product, perhaps not even to the firm's own activities. Hence, it may be able to contract with other firms for the use of its negotiating skills and thereby enhance the *scope* of its investment in developing such skill.

ASCERTAINING THE LEVEL OF TRANSACTION COSTS

Having thus determined wherein the competitive advantage of the firm lies, the corporate strategic decision maker, whether consciously or not, moves to ascertain the transaction costs between separable links in the firm's value-added chain. If these costs are very high, any perceived advantage in externalizing and/or internationalizing a given function or link may be so diluted as to turn the decision around. Transaction cost, of course, refers to

Scale	Scope
These economies arise because of a sunk cost or binding commitment to either:	
Specific factors, which lead to economies of scale *or*	Non-specific factors, which leads to economies of scope
Such as:	Such as:
1-Special purpose machines, bldgs., other capital assets (e.g., transport)	1- General purpose machines, bldgs, other capital assets
2- Specific technical skills (the result of in-hoouse training and/or experience)	2- Generally applicable technical skills, but with experience in working together
3- Specific product and/or process-related development	3- General research and development capability
4- Specialized financial skill (i.e., function or product-related)	4- General financial services and skills
5- Accepted brand names identified with specific services or products	5- Established corporate name (or accepted image) which facilitates market entry and/or access to lower cost capital and other factors (e.g., labor, resources)
6- Specialized marketing skills and services	6- General marketing skills and and services
7- Direct labor (that which embodies specific experience and can be terminated only at considerable expense)	7- Direct labor (which embodies non-specific experience and can be terminated only at considerable cost)
8- Specific information network	8- Multi-purpose information network
	9- Scanning functions and skills
	10- Project evaluation skills of a non-specific sort
	11- Legal capability
	12- Managerial skills
	13- Public affairs skills (i.e., developing and maintaining political leverage, negotiating skills)

Figure 10.3. Differentiation between economies of scale and economies of scope.

the relative ease of communicating the technology relevant to a given function or link.

That is, is it really less costly for the firm to apply a given technology itself (internally) or have it applied by another firm (externally)? In either case, can the activity be undertaken (that is, the technology applied) more cheaply at home or abroad? Continued performance of a given activity by the firm itself domestically may be justified if the transaction cost in dealing with an unrelated party, or with a related party abroad, more than soaks up any potential advantages externalization, or internationalization, might otherwise offer. The point is that, all other things being equal (which they rarely are), internal transaction costs tend to be lower than external transaction costs, domestic transaction costs less than international transaction costs. That is, it is cheaper to deal within the firm itself than without, within the domestic market than outside. Why?

Consider first the externalization/internalization decision in terms of transaction costs. Several reasons for a cost difference in favor of internalized transactions (hierarchical) suggest themselves: (1) people at both ends of the transaction speak the same language in an organizational and technological sense (the stronger the corporate culture, perhaps the greater the advantage in dealing internally), (2) negotiating time and associated legal costs are reduced, though not necessarily, (3) there is minimal risk of nonpayment for commercial reasons (that is, bankruptcy or lack of integrity on the part of the supplier of the activity), (4) no performance bonds or guarantees are involved, and (5) claims based on failure of the technology, product, or know-how to generate the expected results are unlikely. That is, the internal communication of technology is less costly than external communication.

It is also possible to introduce measures so as to reduce the cost of internal transactions still further. First, the firm may undertake activities to accelerate its learning *by maintaining a low employment turnover rate* (and thereby capturing its own experience in technological/organization improvement), *by creating an internal environment such as to facilitate learning and dissemination of that experience* rapidly throughout the organization. The latter can be encouraged by creating a strong *sense of trust and confidence,* maintaining a *low level of threat, encouraging a perception* among employees as to the *fundamental equity* of the way in which the benefits of business success and the costs of reverses are distributed, maintaining *open communication* throughout the organization, developing a strong sense of *participation,* building a *strong organizational culture, restraining direct interpersonal competitition,* and, finally, locating in a culture supporting the values underlying such practices. (This point is discussed later).

But even so, there may be advantages in *external* transactions in that by externalizing certain activities the firm may gain greater flexibility in sourcing. By so doing, it may be able to reduce risk and enable it more easily to exploit a local comparative advantage (that is, to operate at lower cost) other than in countries with which it is familiar. Also, there are at least two important ways in which the cost of external transactions may be reduced. The first is for the firm to maintain a long-standing relationship with the other party, be it supplier, service dispenser, processor, marketer, distributor, financer, or user. A second way, which commonly arises out of the first, is the rotation of individuals between the firm and its external associate. It is not by chance that what is thus described sounds very much like a strategic alliance. In effect, through such measures, the firm may move toward a relationship that encorporates many of the advantages of an internal transaction, as well as of an external transaction.

Now consider the relative cost of domestic and international transactions, whether internal or external. Transaction costs may be exaggerated in the international case by: (1) the imposition of different laws and regulations, (2) the impact of a different culture on understanding, expectations, and motivations, (3) the effect of operating within the restraints of a different monetary system, (4) the impact of international economic and political relations on mutual trust and confidence between parties to a transaction (and, hence, on perceived risk), and (5) the greater difficulty of securing reliable information on markets, costs, individual creditability, and reliability. In order to minimize costs imposed by such elements as these, the firm may either internalize the relevant expertise (experience) or gain access to such by externalizing the function or activity involved—that is, by contracting the experience. In order to externalize functions across borders efficiently, a management must first recognize the signal importance of experience in minimizing the cost. It does not simply rush into unfamiliar territory and expect to carry on with the same efficiency (and cost) as though one were at home on familiar grounds.

In this day of volatile exchange rates, the instability inherent in mounting balance of trade imbalances, political upheaval in many areas, and uncertainty of tomorrow's law and regulation, all of which may impact on a firm's cost and/or earnings in virtually all countries, perhaps it is best to minimize the firm's commitment of its own resources in the bricks and mortar and machines represented by overseas plants that it itself owns, that is, internalized functions via direct foreign investment. Perhaps it may be better to induce someone else to make the investment at risk and to limit the firm's own involvement to the supply of technology and know-how under contract (which may well include quality control and product by-back provisions), a relationship from which the firm may be able to withdraw over a relatively short time with relatively little loss—that is, relative to walking away from its own plant. The firm thereby maintains a position of maximum flexibility, which, given the dynamic nature of the world environment in which it must operate, may be of critical importance in minimizing risk and, hence, cost.

DETERMINING THE COMPARATIVE ADVANTAGES OF COUNTRIES

This discussion leads one into the fourth step in the strategic decision-making process, that of determining the comparative advantage of countries relative to each link in the firm's value-added chain and to the relevant transaction costs. In so doing, one is necessarily alert to: (1) *trading costs* of countries perceived important to the firm (that is, transportation and communication costs, tariffs and nontariff barriers in both directions, taxes, risk and uncertainty of political intervention, and of exchange rate volatility); (2) *factor costs* (the relative costs of capital and labor of various categories actually faced by the firm and of other inputs, such as energy); and (3) *social/ political costs*.

Differences in trading costs are fairly clear and straightforward, though often difficult to tie down, given their ever-changing nature and sometimes novel aspects. The relevance of national differences in factor costs in siting various links in a firm's value-added chain is graphically demonstrated in Figure 10.2, but as in the case of trading costs one can encounter some unfamiliar dimensions. Take labor, for instance. In many countries the cost of legally imposed fringe benefits (social welfare taxes, the annual bonus, payments into a publicly administered pension plan, and termination pay) may be significantly higher than those to which the firm is accustomed. Also, the true cost of labor to the firm includes training, the cost of which is increased by a a high turnover rate and an inadequate publicly supported training system. In addition, domestically based assumptions about the number of employees required may not be justified if productivity is substantially different. Labor law in many countries make labor more of a fixed cost in that termination may be both difficult and costly.

Energy may be a problem. An inadequate supply may force a firm to provide its own source by installing a generating plant, which can be relatively high cost. Local materials, even though relatively cheap, may be of different quality from those to which the firm is accustomed, which may in turn require more rigorous quality control or use of a different production technology. Both energy and materials may be subject to interruption in their flow for a variety of reasons, including ability of local suppliers to secure foreign exchange with which to import needed supplies.

If one is sourcing a product a long way from the market in which it is to be sold, and air freight is not economic, a relatively large inventory may be necessary—including goods in the supply pipeline—because of the time required to ship. Added cost may be the result. And management may be more costly than at first realized if home country nationals (rather than local nationals) are deemed necessary for reasons of control or because of the perceived inadequacy of local managerial talent; the same for technical skills. Setting up an expatriate family abroad can be very costly. Further, sheer distance adds to communications and travel costs, misunderstandings, and the tardy handling of problems. If management is local and does not use the firm's native language fluently, a further cost in translation and delay is incurred. So it goes. Trading and factor costs incurred may be unexpectedly high.

The notion of differing social/political costs is even more complex and yet, relatively little discussed in this context in the literature on strategic decision making. By social/political cost, we refer to costs imposed to a greater degree by some countries than others by reason of the fact that they provide structurally incompatable—that is, high cost—environments for efficient enterprise performance by reason of a number of factors. Some of these are listed below.

1. Highly valued modes of personal behavior may render it difficult for a firm to minimize its internal transaction costs. (Reference is to such modes as exaggerated individualism, aggressiveness, intense interpersonal competition.)
2. The manner of resolving conflicts may require expensive and time-consuming litigation rather than less costly and more rapid modes of conflict resolution, such as informal mediation.
3. The time-horizon forced on decision making, both personal and corporate, may be so limited as to render it virtually impossible for either the firm or its employees to consider other than very short-term costs and benefits.
4. The level of security of persons and property may be such as to require armies of security guards, sophisticated security, and costly insurance.
5. The ability to shift resources may be structurally inhibited, thereby making it costly for a firm to adjust its resources appropriately as markets become increasingly international, as the firm's competitive advantage moves along its value-added chain, and as *national* comparative advantages shift. Examples of such structural inhibitions: union contract provisions, restraints on employee termination or transfer, national trade protection and/or subsidy, performance requirements (re production, employment, local content, export, investment, technology), corporate law (re liquidation, termination, etc.), antitrust or rules of competition, which make difficult mergers and acquisitions.

6. The degree to which a nation holds individual enterprises responsible for unintended injury (possibly even unpredictable by, and unrelated to, the enterprise) to consumers (product liability), to employees (health hazards), and/or to the general public (toxic waste liability, liability for disturbing ecological balances and/or environmental aesthetics). Such liabilities may be relatively high in certain countries, thereby resulting in heightened cost to the firm, either in actual expenditures or the risk of corporate liability, which may or may not be insurable. An example is detailed in this bit of advocacy advertizing by the American International Group, the leading U.S.-based international insurance organization (Fig. 10.4).

7. The level a country feels compelled to an active role in the affairs of other nations may be substantially higher than for another. (Such involvement can lead to politically motivated control of trade, finance, and/or movement of persons—that is, the imposition of sanctions— also to relatively high military investment.)

8. The degree to which a government feels compelled to intervene to control the involvement of foreign business in its society may be relatively great, so likewise the frequency and nature of such intervention and its apparent arbitrariness, that is, unpredictability.

9. The ease with which a society admits and hosts foreign managers and technicians and their families.

The point is, of course, that national social/political behavior and policy can impact on national enterprises, whether locally or foreign-owned, in a number of ways such as to induce relatively high cost. Some of these costs include: repeated entry training and organizational entry, work slowdowns or stoppages, blocked internal communications, slowed introduction of new technology on factory floor or office, heighted litigation and associated legal expenses, major investment in maintaining security, high insurance premiums (or increased risk inherent in self-insurance in the absence of commercial insurance) blocked trade opportunities, high taxes (whether direct or via high interest rates and/or heightened inflation and/or shifts in major foreign exchange rates), and the imposition of a wide variety of government controls. One could go on. The measurement of such social/political costs is made doubly difficult—and hence, costly—by their constantly changing nature.

> In the international marketplace of the future, businesses will be able to locate wherever they want and those companies won't choose places where crime and drug abuse is rampant, families are disintegrating, and employees poorly educated. (U.S. Congresswoman Pat Schroeder, D.-CO., *The News Tribune* [Tacoma], March 25, 1990, p. B-1).

**EXCESSIVE LIABILITY AWARDS MAKE IT TOUGH
FOR U.S. COMPANIES TO COMPETE.**

We are a compassionate society. We want to compensate those who have suffered.

But when our courts expand traditional concepts of liability, causing defendants to pay excessive compensation, we're adding to the costs we all pay for goods and services. We're encouraging companies to stop research and development on new products. And we're even making it harder for U.S. companies to compete overseas.

PAYING A HIDDEN TAX.

In reality, the American system of liability has become the source of a hidden tax on our economy-a tax that can account for as much as 50% of the price paid for a product.

What's worse, it has been estimated that this hidden tax amounts to $80 billion a year-a sum equal to the combined profits of the nation's 200 largest corporations.

Our economic competitor's legal systems do not encourage litigation to the extent we do. Consider, for example, that there are 30 times more lawsuits per capita in the U.S. than in Japan.

Is it any wonder that America is having a tough time competing in overseas markets?

UNCERTAINTY STIFLES ENTERPRISE.

The unpredictability of our liability system is also enormously costly to American competitiveness. For example, in a recent survey of CEO's, the Conference Board found that worry about potential liability lawsuits caused 47% of firms surveyed to discontinue one or more product lines. In addition, 25% stopped certain product research and development, and 39% decided against introducing a new product. Meanwhile, our overseas competitors continue to research and develop new products at an ever increasing pace.

Figure 10.4. *New York Times* (paid advertisement), February 20, 1990, p. C3.

DEVELOPING ADEQUATE FLEXIBILITY IN CORPORATE DECISION MAKING AND ORGANIZATIONAL DESIGN

This last point brings us to the fifth step in the process of corporate strategic decision making, which is perhaps the most difficult of all. This step has to do with designing and developing adequate flexibility, *on an on-going basis,* in decision making and organizational design so as to permit the firm to respond promptly to changes in both its competitive advantages (the externalization–international decision) and in the comparative advantages of countries (the localization decision).

Such flexibility is largely a function of (1) openness of attitude and willingness to learn on the part of decision makers, (2) the level of accumulated, *retained,* and internally communicated experience (that is, organizational learning), and (3) continuing awareness and professional analysis of the cost and benefit of alternative location of functions (that is, internally or externally, domestic or international). Of perhaps lesser import are formal structure and strategy. Although structure and strategy are obviously related to attitude, experience, and awareness and analysis of alternatives, one cannot specify in a *generic* sense the most appropriate structure and strategy of a firm faced with the globalization of its markets. It all depends on the resolution of the five-step, decision-making processes discussed here.

Possibly the only generalization one can make relative to strategy and structure is that a firm's strategic business units (the SBAs) need to be restated in terms of clearly definable, geographically separable functions (applied technology), or links in the firm's value-added chain. An SBU, although certainly enjoying some degree of autonomy in that the management skills relevant to different functions and located in different geographical locations vary, does not necessarily buy and sell to external markets. This definition of an SBU implies accounting innovation for many firms, including a move away from the concept of multiple profit centers to the idea of accountability or responsibility centers, the performance of which is based possibly on engineering or physical input/output efficiency, not in terms of internal financial profit. Therein lies the subject of another study.

Chapter Eleven

International Technology Transfer Literature: Advances in Theory, Empirical Research, and Policy

Tagi Sagafi-nejad

The purpose of this chapter is threefold: (1) to review the relevant literature of the last decade, (2) to assess and categorize that literature by type of contribution made (whether conceptual or empirical), and (3) to identify areas of useful future research. Following this chapter is a rather extensive bibliography that is limited to writings in the English language. *All references in this chapter containing name and/or date in parentheses refer to items listed in the Bibliography.*

INTRODUCTION

In the early 1980s several literature surveys, including Sagafi-nejad and Belfield (1980) and Contractor and Sagafi-nejad (1981) indicated the following themes as being prevalent. First, the literature is multidisciplinary as well as interdisciplinary. Contributions continue to be made from economics (and its various tangents), political science, anthropology, history, international business, international political economy, management, industrial relations, marketing, information systems, law, and other fields. Not only are contributions made by each of those disciplines, but some studies make use of more than one discipline in their research. This is a testimonial to the fact that, like many other arenas of human interaction, technology transfer can best be understood if studied from many angles. Thus Sagafi-nejad and Belfield (1980) classified over 1,500 citations into the following:

1. The role of science and technology in development
2. International technology gap and the New International Economic Order
3. Transnational corporations and technology generation and transfer
4. The anatomy of corporate technology transfer: modes, costs, and management
5. Technology transfer and host countries: appropriateness, dependency, and sovereignty
6. Sectoral analyses: case studies in technology transfer
7. Technology transfer and the host country
8. Regulating technology transfer: control systems and mechanisms.

In their review of some 166 pieces on the subject, Contractor and Sagafi-nejad (1981) concluded that research was needed in the following areas:

1. Theory building, namely integrating the concepts in technology transfer with other bodies of theory, such as foreign direct investment, development, and trade.
2. Measurement. Case studies at the country, industry, and firm levels need to be augmented with cross-section studies, and criteria for measurement would be sufficiently standardized to render comparative analysis possible.
3. Policy research, involving both corporate and national policy instruments. The authors called upon the literature in R&D management and planning to be extended to the analysis of multinational technology transfer. In addition, they urged that the questions of "optimization" of the mode, price, and territorial coverage from the supplier firms' viewpoint, as well as the optimal regulatory and instutitutional development from the host country perspective, needed further research.

To what extent has the research of the 1980s added to our knowledge in these areas? It has certainly exploded in volume, evidenced by the sizeable, yet undoubtedly incomplete, bibliography following this chapter. Two general themes stand out. First, this expansive literature is amorphous, with roots in many disciplines and with an increasingly interdisciplinary flavor. Second, due to this heterogeneity, no single overarching theory has yet emerged. A grand theory may not be feasible, or even desirable, but a conceptual umbrella can be devised within which the current state of the art can be understood and interrelated. Such an umbrella began to emerge during the years, implicit in many of the works reviewed here.

This chapter makes a modest attempt, by fleshing out this framework, to help organize the current state of knowledge in international technology transfer. We begin by noting the major themes in the literature of the past decade. We then propose the conceptual anchor within which the literature can be better understood.

MAJOR THEMES

The Changing Technological Balance of Power

Competitiveness became a buzz word in both academia and policymaking circles in the 1980s, spawning a vast literature. Historians and futurologists alike saw either gloom and doom or a bright future for America's competitive position. Governments were called to action (e.g., PCIC 1985; Young 1988). Both the pessimists and the optimists, joined by such luminaries in the policy arena as Blumenthal (1988) and Wriston (1988/89), agree on the critical role of technology in maintaining or attaining global dominance, thus confirming earlier theory (Jones 1971; Williams 1972).

As the international system evolves from hegemonic to multipolar and pluralistic, we note that the international flow of technology is increasingly multidirectional. We have the literature on competitiveness (Baily and Chakrabarti 1985; Cohen and Zysman 1987; Davidson 1984; Gamota and Friedman 1988; Giles and Brooks 1987; Gunn 1987; Hatsopoulos et al. 1988; Lawrence 1984; Milke and Weston 1988; National Research Council 1984; PCIC 1985; Sarr 1988; Scott and Lodge 1985). Here both evidence and opinion are mixed, ranging from "declinists" (Kennedy 1987; Lodge 1984); to the optimists (Cetron and Davies 1989; Nye 1990).

An underlying theme common to both camps, as well as to those in between, is that competitiveness at the national and corporate levels is determined by technology transfer and related issues such as R&D management. At the national level, these are science and technology, technology export controls, tax and other incentives, and similar macroeconomic considerations. At the level of the firm, critical variables are corporate strategy, commitment to technology, R&D management, structure of the industry, position of the firm within the industry, motives for market entry strategy, and similar factors. The implications of this literature for technology transfer are manyfold. First, the globalization of technology (Giles and Brooks 1987; Grunwald and Flamm 1985; Muroyama and Stever 1988) implies a global diffusion of technology that brings about greater competition among suppliers of technology (Gunn 1987) with obvious positive outcomes for buyers. A consequence is that new companies from Japan and the NICs join the international market as purveyors of technology. In contrast, globalization of technology also has been accompanied by a tendency toward bigness and consolidation. The wave of strategic alliances (Contractor and Lorange 1988; Ohmae 1989; Perlmutter and Heenan 1986; UNCTC 1987, 1988) is one sign of global consolidation and concentration in certain high technology fields.

Contrary signs can be seen in other ways, too. Not only have Japan (Henkoff 1989; Gamota and Friedman 1988) and NICs become important players it is conceivable that, in certain technological niches, newly liberated economices of Central and Eastern Europe will become more active players in technology trade. There are already signs that such a move is afoot; see Kiser (1989) and Stipp (1990).

Despite this increase in the multidirectionality of the flow of technology, we would be remiss in assuming that the dominant role of U.S. multinationals as powerhouses for the generation and transfer of technology is at an end. Table 11.1 shows worldwide technological "balance of payments" for selected countries and years between 1970 and 1985. Note that the U.S. payments-to-receipts ratio has in fact increased from 6.5 to 7.3, and it remains the largest beneficiary of royalty payments, and at an accelerated tempo.

Table 11.1
Worldwide Payments and Receipts of Royalties and Licensing Fees, 1970–85

	U.S.	Japan	U.K.	West Germany	France	Italy	Developing Countries
Payments ($ billions)							
1970	0.2	0.4	NA	0.4	NA	NA	0.6
1975	0.5	0.7	0.5	0.8	0.5	0.6	1.2
1980	0.7	1.3	0.9	1.4	1.0	1.2	2.3
1985	0.8	2.4	0.9	1.2	1.0	1.5	NA
Receipts ($ billions)							
1970	1.3	0.1	NA	0.2	NA	NA	NA
1975	2.6	0.1	0.6	0.3	0.2	0.3	NA
1980	5.0	0.4	1.1	0.6	0.5	0.8	NA
1985	5.8	0.7	1.0	0.6	0.5	1.0	NA
Net (Receipts less Payments)							
1970	1.1	−0.3	NA	−0.2	NA	NA	−0.6[a]
1975	2.1	−0.6	0.1	−0.5	−0.3	−0.3	−1.2[a]
1980	4.3	−0.9	0.2	−0.8	−0.5	−0.4	−2.3[a]
1985	5	−1.7	0.1	−0.6	−0.5	−0.5	NA
Ratio (Receipts/Payments)							
1970	6.5	0.3	NA	0.5	NA	NA	NA
1975	5.2	0.1	1.2	0.4	0.4	0.5	NA
1980	7.1	0.3	1.2	0.4	0.5	0.7	NA
1985	7.3	0.3	1.1	0.5	0.5	0.7	NA

[a]Data for LDC receipts are unavailable, but the amounts are presumed to be neglibible.
NA = not available.

Source: Benjamin Gomes-Casseres, "Note on Global Technology Flows," Harvard Business School, 1989, mimeo.

Costs and Compensation Issues

Perhaps the least documented and understood aspects of the field, questions of cost and compensation, have confounded researchers for several reasons. First, it is the most secretive aspect of technology transfer. Neither companies nor governments are too eager to divulge compensation data. Second, even where data has been reported, such as Contractor (1981), Kopits (1976) and the handful of other studies, they are not comparable across industries and countries. In a case study consisting of a 50-plus sample of technology transfer agreements with foreign firms in Iran, Sagafi-nejad and Magee (1982) found some industry variations in royalty rates, but this data cannot readily be extended to other countries for lack of comparable statistics. Third, the total compensation package can be impossible to determine, let alone compare, because of the variety and complexity of the various forms of payment. Finally, there are costs associated with the development and transfer of technology that are often hard for the firm to document and substantiate. And finally, when technology transfer is seen in the broader context, larger trade-offs are involved both for the firm and the importing country. How critical is this technology to the competitiveness of the targeted industry? How badly does the recipient firm want it? Conversely, how can this technology transfer be used as an element of an overall entry strategy by the MNC? If it is a prelude to a broader or deeper involvement, where supplies of machinery, parts, and associated product may be involved, the potential gains must be factored in. None of these issues have been studied in sufficient detail to allow general conclusions. The best that can be said at this juncture is that, as a transfer variable, price is critical to the whole process.

The Role of Patents

It is axiomatic that the protection of proprietary technology lies at the heart of global competition. How firms generate, protect, and use technology determines their competitive position, and how national policies nurture or accommodate this strategy, in turn, dictates national competitiveness. As the Uruguay Round of Multilateral Trade Negotiations draws to a close, it is more important than ever to unravel the means by which multinationals and other firms attempt to control the world's technology and to understand the conflicting and complementary interests of firms and nations.

Ever since the Paris Union for the protection of intellectual property (WIPO's grandfather) was established in 1883, debate on patents and intellectual property has continued unabated. There are good reasons for this. Countries need policies that reward innovative activity, and innovators are motivated by economic incentives; hence the twin objectives of maximizing society's technological vitality and minimizing the potential harm from monopolistic behavior of firms. Patents play a key role in this process because, inter alia, they provide the legal barriers to competitive imitation, thus shielding the innovator long enough to gain from dynamic efficiency.

Because the protection of intellectual property rights constitutes a key agenda item in the current round of GATT negotiations and because the need to generate rigorous empirical data to help guide policy and future research is rather urgent, Bertin and Wyatt (1988) present empirical data from American, European, and Japanese multinationals regarding the role of technology in corporate strategy, patents, and other forms of protecting proprietary technology, and nationality and industry differences in practice among companies. What are the Schumpeterian versus Ricardian "rent" considerations? Does the patent system promote technological dynamism by granting temporary monopoly to the owner, or does it help bring about the eventual diffusion of technology as a "public good"? How does the appropriability of technology help the firm in its strategy to extend its own production activities abroad through internalization rather than licensing or other externalized modes? What is the relative efficacy of the patent system in contradistinction to other forms of protecting proprietary knowledge? Are there interindustry and intercountry differences? And what about the juxtaposition of the competitive forces between firms and nations? Bertin and Wyatt (1988) reiterate the MNCs' need for global protection of technology and describe the methods of protection, based on empirical data. Their primary tenet is that:

> Technology, especially that which operates in-house, appears to be the major source of competitive advantage for multinationals; and . . . the use of industrial property rights, notably patents, is a significant means of consolidating technological advantage. (pp. 23–24)

The Regulatory Environment

The 1980s witnessed a liberalization in the regulatory environment for transfer of technology. In the north-south context, this was due to the dire predicament in which most LDCs found themselves because of the debt burden. According to an UNCTAD study (1988), although the global flow of technology accelerated in the 1980s, and at a time when global competition for development and sale of technology intensified between firms in the United States, Western Europe, and Japan, the developing countries' share of technology flows remained stagnant. Many LDCs were basically kept out of the loop due to their declining economic performance. More fundamentally, developing countries perceived a declining interest on the part of the MNCs to transfer technology under conditions they considered suboptimal. The breakdown and eventual dismantling of central planning in East and Central Europe, which began in the fall of 1989, served only to hasten this drive. In the advanced countries, governments were increasingly disinclined to interfere with market forces.

Nevertheless, there still remain a plethora of controls that could affect the flow of technology. These are not all intended to supplant the "market solution" but to eliminate its imperfections and to maximize the benefits and minimize the disadvantages. Varying in rigidity and enforceability, controls emanate from the following: corporate self-control, and national, bilateral, and multilateral sources. At the corporate level, some of the concern shifted to the corporate liability issues, as MNCs found their presence more desired and their obligations more problematic. The Bhopal catastrophe in India in December 1984 brought the matter to worldwide attention. The Union Carbide Corporation found itself having to account for the manner in which it had transferred the technology for the manufacture of chemicals and associated technologies for maintenance and crisis management to its joint venture plant in Bhopal. At the same time that companies felt harassed by peripheral markets, they found their most strategic asset even more desirable and hence more worthy of protection (see Reidenberd 1988; Robinson 1988). Technology management thus moved from the basement to the executive suite, as more firms sought to articulate their technology transfer policy as part of their overall strategic management. Self-control became not only a matter of damage control but also one of maximizing the benefits through externalized as well as internalized use of technology on a global scale.

At the national level, the legislative and administrative environment took on an increasingly liberal and accommodating tone. Mexico, a pioneer in setting up technology transfer laws and a registry, liberalized its legal apparatus in late 1989 in hopes of attracting more technology and foreign investment, and as part of its own liberalization drive. Similar efforts were underway in LDCs as well as in Central and Eastern Europe and the Soviet Union.

Under UNCTAD auspices, regular exchanges of experience did occur among the technology transfer registries, who seemed to concur that

1. Monitoring technology transfer is a very complicated, skilled labor-intensive and information-intensive operation
2. Technology transfer registries in developing countries should initiate and/or expand monitoring activities wherever feasible
3. The main objective of the regulatory agencies created in many developing countries for the monitoring analysis and control of technology transfer agreements between domestic and foreign parties have been to:

- stimulate flows of technology to preferential areas according to national development priorities
- increase the bargaining power of domestic companies and reduce the direct and indirect costs of imported technology
- improve the conditions for an effective absorption and assimilation of know-how
- support the development of domestic scientific and technological capabilities (UNCTAD 1984, pp. 2–3).

Despite the rhetoric, as noted earlier, the regulatory environment continued to liberalize, albeit at varying and generally slow pace (see UNCTAD 1988a, b; UNCTC 1987).

Although as "international regime" for the transfer of technology has not yet emerged, one can identify the contours of such a system. The key institutional players are likely to be the General Agreement on Tariffs and Trade (GATT), the World Intellectual Property Organization (WIPO), the United Nations Conference on Trade and Development (UNCTAD), and the United Nations Centre on Transnational Corporations (UNCTC). GATT took up the discussion of technology transfer rather inadvertently, when the issues of intellectual property and trade-related investment measures became part of the Uruguay Round in 1986. It is expected that the new round will have an impact on technology transfer. WIPO, meanwhile, has been groping with the question of how best to safeguard the interests on all the interested parties in its new round of revisions of the Paris Union (last revised in 1967). It continues to develop detailed "model laws" for adoption by nation states. UNCTAD, long a battleground between the LDCs (Group of 77) and advanced market economies, has been at work, if only at a glacial pace, on two fronts. The set of rules on Restrictive Business Practices (RSBs), passed unanimously but without a formal vote in Geneva in April 1980 (Sagafi-nejad 1989; UNCTAD 1981), was negotiated as a multilateral instrument to prevent TNCs from exercising a dominant market power, especially in their relations with LDCs. The primary reason for the apparent consent by Western countries was that these rules were consonant with much of their own anticompetition laws. Similar collective attempts by the G-77 to arrive at legally binding and enforceable codes of conduct for technology transfer, meanwhile, have been stalled for some 15 years, and the debate will not go away. It does not seem to be getting anywhere either. Meanwhile, there appears to be some evidence suggesting that RBPs are occurring or have been noted in the recent past. UNCTAD monitors these developments, assisted by governmental monitoring agencies whose implementation of these instruments varies among participating countries.

Elsewhere, there are industry and country case studies that deal with RBPs in some fashion. In his 1977 study of 35 joint ventures in Iran, Rafii (1984) found evidence of delimiting clauses expressly restricting the scope of local manufacture, tied-purchase clauses, quality control, and supply restrictions, prohibitions against the use of third party know-how, and deletion allowances where components prices are based on marginal cost as a disincentive to local purchases. He found, moreover, that domestic content was adversely affected by foreign control.

Technology Transfer from the Third World

The ability of a country to export technology has been taken as a good indicator of technological mastery or absorption (Dahlman and Westphal 1981; Lall 1984; Sagafi-nejad 1979). The 1980s witnessed an acceleration in the pace of such exports from selected developing countries, primarily NICs. Whereas this points to a widening gap within the Third World, it also means that these countries have reached a degree of technological competitiveness through enhanced technological absorptive capacity that must be increased before a country can join the ranks of technology exporters. Furthermore, today's exporters are yesterday's aggressive technology importers. To wit: Japan. Countries such as Taiwan, South Korea, India, Mexico, and Brazil thus became technology exporters during the last decade. Lall (1984) reports case studies of Hong Kong, Taiwan, Republic of Korea, India, Egypt, Mexico, Argentina, and Latin America. He observes that, with the exception of Egypt, many of the more advanced developing countries are becoming serious exporters of standardized technologies and person-embodied know-how. Even in the case of Egypt, Sagafi-nejad (1984) found that Egyptian technology exports were more notable than commonly perceived. Many of the transfers were motivated primarily by political and cultural factors, often consisted of person-embodied technology, and were done by large firms already involved in exporting. Other case studies in Lall (1984) point to much greater technology export performance, notably by India, Brazil, South Korea, and Taiwan.

Noncommercial Transfers

The literature has focused almost exclusively on commercial transactions among firms, to the relative neglect of two rather important forms of transfer, namely transfers that are state-influenced and those that are illicit. The first includes all the bilateral and multilateral scientific exchange programs. Due to the foreign policy considerations that underlie these exchanges, the president of the United States has been mandated since 1979 to submit an annual report to Congress titled "Science, Technology and American Diplomacy" (see President's Report to the Congress annual). These reports cover the so-called bilateral S&T agreements as well as multilateral work on science and technology. Sagafi-nejad (1987) identified an extensive range of flows in the case of the United States, admittedly the most active country in this field.

All such efforts are, in fact, incidents in the transfer of technology, at times with commercial undertones if not direct connections, but are neglected thus far in the literature.

A CONCEPTUAL FRAMEWORK

Based on a screening of the literature, we suggest that understanding international transfer of technology can be fruitfully moved forward by examining four sets or clusters of variables: technology, transfer, organizational, and environmental. *Technology* contains those variables that define the properties of the technology being transferred, its complexity, life cycle, and other characteristics. *Transfer* includes issues pertaining to the transfer arrangement between the supplier and recipient of technology, including the modes, methods, and modalities of the arrangement and whether it is part of a joint venture arrangement, as well as price and compensation and other details of the agreeement. *Organizational* variables define the supplier and recipient firms, their needs, technological capabilities, relative bargaining power, and other "organizational profile" characteristics. *Environmental* variables are in the home and host countries of the two firms, most notably the level of development and technological absorptive capacity of the host country. The interaction of these four clusters of variables help us understand whether transfer will take place, the mode of transfer (arm's length "externalized" vs. through a wholly owned subsidiary or another "internalized" form), price and other details of the transaction, and its impact on the host and home countries as well as the supplying and receiving firms.

The above conceptualization can also help managers and policymakers with their tasks while enabling researchers to get a tighter conceptual grip on the amorphous subject of technology transfer. First we examine the many changes in the international environment within which technology is transferred. This framework can be graphically illustrated as in Figure 11.1.

Technology Variables

To understand the TT process, a typology of technologies is needed. How complex, how capital-intensive, how proprietary, how far along the product life cycle . . . ? These all help define the nature of the technology being transferred. Certain technologies are more easily transferred, adopted, and

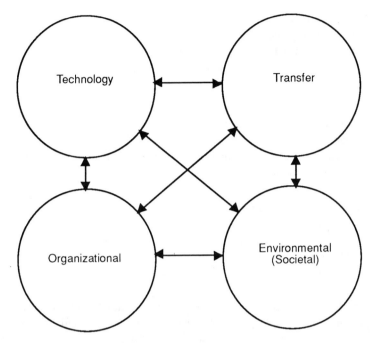

Figure 11.1. Interdependent variables in the technology transfer process.

diffused. The "technology" variables are those that help differentiate technologies from one another, be it along industry lines, complexity, life cycle, R&D intensity, or another attribute. Several theories of international business, including Buckley and Casson (1976), Vernon (1967 and 1970), Rosenberg and Frischtak (1985), and Cantwell (1989) place technology at the core of their conceptual framework. Technology is also explicit in the internalization theory (Dunning 1978 and 1983; Buckley and Casson 1976), internalization (Rugman 1983) and appropriability (Magee 1976). As Kogut and Zander (1988) have noted, competition among firms and nations is driven by the differences in the amount of know-how and information they each possess.

Terms of the Transfer Arrangement

We refer to the terms and conditions under which transfer takes place as transfer variables. They define the mode and method of transfer and the contractual relationship between the supplier and the recipient of technology. Various modes of transfer are

- Turnkey projects
- Wholly owned subsidiaries
- Joint ventures with managerial control
- Joint venture minority (no control)
- Arm's length licensing.

Each of the above modes places a different set of conditions and constraints on the transfer process.

The method by which technology is transferred also helps determine its effects. These methods include

- Planning and feasibility studies
- Product design
- Machinery and equipment
- Exchange of personnel
- On-the-job training.

In short, transfer variables are constraints, conditions, and parameters set forth in the contractual agreement between the supplier and the recipient. The TT contract can contain various RBPs, for instance, that would limit the freedom latitude of the recipient in its search for local subcontractors, vendors, and suppliers. Other restrictions placed on the agreement could have similar negative effects.

On the positive side, explicit provisions for local training, gradual increase in local content, or a promise of refraining from certain restrictive practices can only help facilitate the effectiveness of the technology transfer process. Studies that highlight the transfer variables are exemplified by Thunman (1988), whose work deals with the relationship between Swedish technology supplier firms and their Indian recipients (see especially pp. 106–7 and passim).

The Organizational Dimension

As stated earlier, transfer takes a variety of organizational forms. Studies focusing on this dimension include Westney (1990 this volume), Morgan (1990 this volume), and Stobaugh and Wells (1984), and Yeh and Sagafi-nejad (1987). Westney, as demonstrated in Chapter Nine of this book, uses the historical case study of Japan's quest for importing organizational technologies during the Meiji restoration in the 1870s. Here the primary focus is on organization qua technology, thus highlighting the role of the organizational dimension. Stobaugh's and Wells' study of technology transfer in the petrochemical industry, on the other hand, addresses another aspect of the same issue, namely the channels within which technology are transferred. These and similar studies complement and support Dunning's contention (see Sagafi-nejad 1981) that the organizational modalities for transfer of technology (and more generally resources) will vary depending on market imperfections and costs, and that on certain occasions the "market solution" will prevail, whereas at other times a firm would "internalize" these transfers.

The various attributes of both supplier and recipient firms thus influence the nature and intensity of impact of transferred technology. On the supply side, they include a distinction between transnationals and nontransnationals and other classifications of these organizations along such dimensions as national origin. Similarly, recipient firms should be classified as to whether they are traditional, transitional, or modern in structure, whether they are privately or publicly held, the characteristics of their top management, the extent to which management and ownership overlap, and other relevant organizational factors as well as those pertaining to the supplying firms.

The accelerated tempo of strategic alliances in the 1980s can also be seen as another organizational form for technology transfer. Here the organizational structures of the firms in the alliance are among the dominant factors. Studies by Harrington (1988) have shown that complementary missions, resource capabilities, and other organizational characteristics create more of a "strategic fit," thereby contributing to the success of the alliance. Her results suggest that "ventures last longer between partners of similar culture, asset sizes, and venturing experience levels . . . [and] when their activities are related." (Harrigan in Contractor and Lorange 1988, p. 225).

In their study of industrial property management by multinationals, Bertin and Wyatt (1988) describe the organizational arrangements used in technology management and transfer. Empirical evidence shows that the methods for protecting intellectual property vary by industry and by national origin.

There are notable company and country differences. Japanese firms, for instance, tend to rely more heavily on patenting than their U.S. and Western European counterparts. Whether this is due to the differences in corporate strategy or legal environment remains unanswered. Do Japanese firms rely on "patent flooding" because of the nature of the Japanese patent system, or is this likely to be the case as Japanese sheer technological prowess? Further research is needed here.

The Environmental (Societal) Variables

The success or effectiveness of the technology transfer process depends, inter alia, on the macroeconomic and other societal and environmental forces and contexts within which it takes place. Foremost among these is the infrastructural capabilities of the host country, technologically and otherwise.

Societal variables are macro-level contextual and environmental variables that help define a society's factor endowments, capabilities, and constraints. Studies that have brought these variables to center stage include Dahlman and Westphal (1981), Lucas and Freedman (1983), the OECD (1981), OTA (1984) study of technology transfer to the Middle East and Succar (1987). These and other studies yield the following list of salient societal or contextual variables:

1. Economic:
   GNP size and growth rate
   GNP per capita and growth rate
   Rate and size of savings and capital formation
   Market size (demand)
   Relative size of manufacturing sector
   Price mechanism—factor proportions and factor distortions
   Labor force—size and quality.
2. Cultural:
   Cultural propensity toward innovation
   Individual's attitudes toward science and technology
   Attitudes toward foreign investment and technology
   Achievement orientation, motivation, risk-taking propensity
   Attitude toward work, authority, discipline
   Rising expectations and management of social tension in different cultures.
3. Political:
   Structure of the political system
   Political elite
   Policy toward foreign investment (political risk factors).
4. Administrative-legal:
   Explicit science and technology policies
   Developmental institutions and mechanisms
   Administrative organs for policymaking and for implementation
   Laws and regulations on foreign investment, science, and technology
   Patent and industrial property laws
   Social legislation (labor laws, welfare measures, environment regulations).
5. Infrastructural:
   Scientific and technological centers
   Manpower training institutions
   Institutions for promotion of transfer and/or development of technology
   Financial institutions.

This list can be expanded and further disaggregated. Our intent in presenting it here is to illustrate the type of disaggregation that needs to be undertaken in order better to understand the societal and macro-level factors that influence the effectiveness of imported technology. Several studies including Goulet (1977), Kedia and Bhaget (1988) emphasize two aspects: (1) cultural variables affect technology transfer, but (2) technology is not value-neutral. Its value dimension, too, is critical.

Cross-impact

In addition to the direct impact of each variable or cluster, there is also ample evidence in the literature that shows the cross-impact phenomenon, namely, the impact of one variable, or group of variables, on another. This is to recognize that the four clusters of variables (i.e., technology, transfer, organizational, and societal) impact not only the transfer process as a whole but one another as well. Thus a large, modern local firm whose owners/ managers are well connected in the elite structure is more likely to reach favorable terms of transfer than firms without those attributes. Here we see cross-impact of "organizational" and "transfer" variables. Many more instances can be conceived.

For instance, Kogut and Zander (1988) have addressed some of the interactions between organizational change and technology transfer. They observe that transfer and organizational change are related and that effective transfer is a function of the recipient's absorptive capactiy and that proprietary know-how is best safeguarded through secrecy. This corroborates the internalization notion that the more complex the technology, the more likely for it to be transferred intrafirm. This is cofirmed by the literature in strategic alliances, including those reported in Contractor and Lorange (1988) and Mowery (1989).

This points out the systematic totality of the four clusters. Not only does each variable in a given cluster have an impact on the transfer process, it is likely to affect other variables in its own and/or other clusters as well.

In their study of the transfer of aerospace technology from the United States to Japan (1955–66), Hall and Johnson (1970) found Japan's high technological absorptive capacity, i.e., the existence of a sizeable network of companies with firm-specific, system-specific, and general technological capabilities, to be a key determinant in the successful transfer of the F-104J and other aerospace technologies to Japan. Similarly, where imported technology has failed to take hold and be diffused, lack of adequate absorptive capacity is as much to be blamed as inappropriate technology. Indeed the two may amount to the same thing.

The OTA study of technology transfer to the Middle East (1984) found

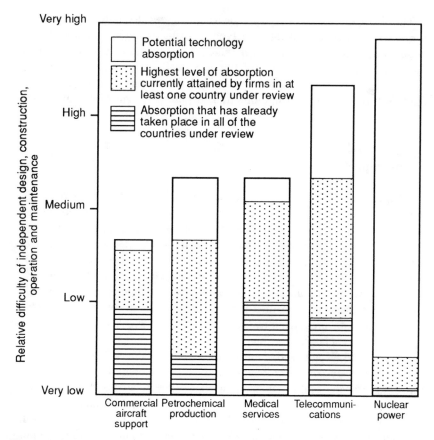

Figure 11.2. Technology absorption in the Middle East, 1984.

different degrees of technology absorption among different industries (see Figure 11.2). Studies of technology exports from developing countries (e.g., cases of technology exports from South Korea, Taiwan, Argentina, India, and Egypt in Lall, 1984) also indicate that technological mastery, and thus the ability to export it, varies among developing countries. Thus the strategy of the firm, too, will be dictated in part by the technological capability of the recipient firm as well as its ability to draw upon the pool of available technological reservoir in the host country.

CONCLUSION

Technology transfer literature is exploding. Several generalization can be made:

Technology transfer is multidisciplinary. Many scholars from economics, political science, history, organization behavior, finance, accounting marketing, anthropology, and law have made significant contributions. Some have even gone so far as to suggest their own theory (Barton et al. 1989, in law; Robinson in management, Dunning, Teece and Contractor in economics, etc. Many of these are hybrid theories that combine tools and concepts of several fields. Yet, ironically, many studies have not been cumulative.

Technology transfer is multidirectional. As modernization spreads and technological change diffuses across countries, the international markets for many technologies become less oligopolistic. More sources of supply become available, even from hitherto unusual places. Kiser (1989) documents nearly 50 examples of East bloc technologies that have been licensed to U.S., Japanese, or Eastern European firms since the 1960s. As the revolutionary changes in the Soviet Union unfold, we are likely to see more attempts on the part of the Soviets to "internationalize the conduct of scientific research and the development of its application," in the words of an Arthur D. Little official who is helping to commercialize technologies developed by scientists at the Soviet Academy of Sciences (see Stipp 1990).

Areas of further research. Despite the proliferation, there are many areas that need further exploring, including:

1. Technology transfer in the services industry. Despite the proliferation of research, there are still many aspects of technology transfer yet to be fully researched. The services sector is posing many new questions as the world economy moves toward services and as the technology involved in services becomes more critical to global competition. Take the many acquisitions of U.S. financial firms by the Japanese, where hundreds of millions of dollars are paid to acquire the technologies developed by brokerage firms, merchant bankers, and other service companies. These technologies include developing and offering to the financial community on a global scale such "products" as program trading, stock index arbitrage, and international portfolio trading. Nomura Securities paid several hundred million dollars in 1989 to acquire Wall Street firms whose primary assets were not physical (desks, computers, buildings, etc.) but the technological capabilities of its staff and its proprietary computer software in areas targeted by Nomura. We know little about the nature of these soft, firm-specific technologies, how they are transferred, at what price, and with what consequences to the suppliers and recipients.

2. Managerial dilemmas. As theory building proceeds in international business, research needs to focus on the tradeoffs between various modes

of transfer, given the full gamut of variables that are emerging. Dilemmas include strategic considerations in the choice of channel, price, royalty duration, and the like. These are all the more critical as the arena of choice expands for both the buyers and sellers in the global technological marketplace.

BIBLIOGRAPHY

Baranson, Jack (1978) *Technology and the Multinationals.* Lexington, MA: Lexington Books.

Baily, Martin Neil, and Alok K. Chakrabarti (1985) "Innovation and U.S. Competitiveness," *The Brookings Review* 4, no. 1 (Fall), 14–21.

Barton, John H., Robert B. Dellenbach, and Paul Kuruk (1989) "Toward a Theory of Technology Licensing," *Stanford Journal of International Law* 25 (1): 195–229.

Baumol, William J. (1989) *Productivity and American Leadership: The Long View.* Cambridge, MA: MIT Press.

Bertin, Gilles Y., and Sally Wyatt (1988) *Multinationals and Industrial Property: The Control of the World's Technology.* Wheatsheaf, England: Haverster.

Bertsch, Gary, and John McIntyre, eds. (1983) *National Security and Technology Transfer.* Boulder, CO: Westview Press.

Blumenthal, W. Michael (1988) "The World Economy and Technological Change," *Foreign Affairs* 66 (3): 529–50.

Buckley, Peter J., and Mark Casson (1976) *The Future of the Multinational Enterprise.* London: Macmillan.

Cantwell, John (1989) *Technological Innovation and Multinational Corporations.* Oxford: Blackwell.

Cetron, Marvin, and Owen Davies (1989) *The American Renaissance: Our Life at the Turn of the 21st Century.* New York: St Martin's Press.

Chacko, George K. (1986) "International Technology Transfer for Improved Production Functions," *Engineering Costs & Production Economics* 10, no. 3 (September), 245–52.

Chandler, Alfred D. (1962) *Strategy and Structure.* Cambridge, MA: MIT Press.

Chen, Edward K. Y. (1983) *Multinational Corporations, Technology and Employment.* New York: St. Martin's Press.

Cohen, Stephen S., and John Zysman (1987) *Manufacturing Matters: The Myth of the Post-Industrial Economy.* New York: Basic Books.

Contractor, Farok (1981) *International Technology Licensing: Compensation, Costs and Negotiations.* Lexington, MA: Lexington Books.

——— (1985) *Licensing in International Strategy: A Guide for Planning and Negotiations.* Westport, CT: Quorum Books, Greenwood Press.

Contractor, Farok, and Peter Lorange, eds. (1988) *Cooperative Strategies in International Business.* Lexington, MA: Lexington Books.

Contractor, Farok, and Tagi Sagafi-nejad (1981) "International Technology Transfer: Major Issues and Policy Responses," in *Journal of International Business Studies* (Fall), pp. 113–35, reprinted in William Dymsza and Robert Vambery, eds., *International Business Knowledge*. New York: Praeger, 1987, pp. 515–37.

Cooper, Charles (1972) "Science, Technology, and Production in the Underdeveloped Countries," *Journal of Development Studies*, 9, no. 1 (October).

Dahlman, Carl J., and Larry E. Westphal, (1981) "The Meaning of Technological Mastery in Relation to Transfer of Technology", *The Annals* of the American Academy of Political and Social Science (November), pp. 12–16.

Davidson, William H. (1984) *Amazing Race*. New York: John Wiley & Sons.

Davidson, William H., and Donald G. McFetridge (1985) "Key Characteristics in the Choice of International Technology Transfer Mode," in *Journal of International Business Studies* 16, no. 2 (Summer), 5–21.

Derakhshani, Shidan (1986) "Negotiating transfer of technology agreements," *Finance and Development* 23, no. 4 (December), 42–44.

Dertouzos, Michael L., Richard K. Lester, Robert Solow, and the MIT Commission on Industrial Productivity (1989) *Made in America: Regaining the Productive Edge*. Cambridge, MA: MIT Press.

Dunning, John H. (1981) "Alternative Channels and Modes of International Resource Transmission," in Sagafi-nejad et al., 1981, pp. 3–26.

——— (1983) "Market Power of the Firm and International Transfer of Technology," *International Journal of Industrial Organization*, 1: 333–51.

Engholm, Christopher (1989) *China Venture: America's Corporate Encounter with the People's Republic of China*. Glenview, IL: Scott, Foresman.

Ferguson, Charles H. (1989) "America's High-Tech Decline," *Foreign Policy* no. 74 (Spring), pp. 123–44.

Frame, J. Davidson (1983) *International Business and Global Technology*. Lexington, MA: Lexington Books.

Gamota, George, and Wendy Friedman (1988) *Gaining Grounds: Japan's Strides in Science & Technology*. Cambridge: Ballinger.

Giles, Bruce R., and Harvey Brooks, ed. (1987) *Technology and Global Industry: Companies and Nations in the World Economy*. Washington, DC: National Academy Press.

Gilpin, Robert (1987) *The Political Economy of International Relations*. Princeton, NJ: Princeton University Press.

Goulet, Dennis (1977) *The Uncertain Promise: Value Conflicts in Technology Transfer*. New York: IDOC/North America.

Grunwald, Joseph, and Kenneth Flamm (1985) *The Global Factory: Foreign Assembly in International Trade*. Washington, DC: Brookings.

Gunn, Thomas G. (1987) *Manufacturing for Competitive Advantage: Becoming a World Class Manufacturer*. Cambridge: Ballinger.

Hall, G. R., and R. E. Johnson (1970) "Transfers of United States Aerospace Technology to Japan," in Vernon (1970), pp. 305–58.

Harringan, Kathryn Rudie (1988) "Strategic Alliances and Partner Asymmetries," in Contractor and Lorange, (1988), pp. 205–26.

Hatsopoulos, George N., Paul R. Krugman, and L. H. Summers (1988) "U.S. Competitiveness: Beyond the Trade Deficit," *Science* 241, no. 4863 (July 15), 299–307.

Heilbroner, Robert (1988) "Is America Falling Behind?" *American Heritage* (September/October), pp. 95–100.

Henkoff, Ronald (1989) "What Motorola Learns from Japan," *Fortune* (April 24), pp. 157–68.

Heston, Alan W., and Howard Pack, special eds. (1981) "Technology Transfer: New Issues, New Analysis," *The Annals* of the American Academy of Political and Social Science (November).

Hirschman, Albert O. (1958) *The Strategies of Economic Development*. New Haven: Yale University Press.

Hughes, Thomas P. (1981) "Transfer and Style: A Historical Account," in Sagafinejad, et al., 1981.

Huntington, Samuel P. (1988–89) "The US—Decline or Renewal?" *Foreign Affairs* (Winter), pp. 76–97.

Hymer, Stephen (1972) "The Multinational Corporaton and the Law of Uneven Development," in Jagdish N. Bhagwati, ed., *Economics and the World Order from the 1970's to the 1990's* London: McMillan.

Jones, Graham (1971) *The Role of Science and Technology in Economic Development*. London: Oxford University Press.

Jorgenson, Dale W., Masahiro Kuroda, and Mieko Nishimizu (1986) "Japan-US Industry-level Productivity Comparisons 1960–1979." Cambridge, MA: Harvard Institute of Economic Research.

Katz, Michael L., and Carl Shapiro (1986) "How to License Intangible Property," *Quarterly Journal of Economics* (August), pp. 567–89.

Kedia, Ben L., and Rabi S. Bhagat (1988) "Cultural Constraints on Transfer of Technology Across Nations: Implications for Research in International and Comparative Management," *Academy of Management Review* 13, no. 4 (October), 559–71.

Kennedy, Paul (1987) *The Rise and Fall of the Great Powers: Change and Military Conflict from 1500 to 2000*. New York: Random House.

Kiser, John W. (1989) *Communist Entrepreneurs: The Unknown Innovators in the Global Economy*. London: Franklin Watts.

Kogut, Bruce, and Udo Zander (1988) "Knowledge of the Firm and the Internationalization of Technology," *Working Paper*, R. J. Jones Center. Philadelphia: The Wharton School, University of Pennsylvania.

Kopits, George (1976) "Intrafirm Royalties Crossing Frontiers and Transfer Pricing Behavior," *The Economic Journal* (December), pp. 791–805.

Kotkin, Joel, and Y. Kishimoto (1988) *The Third Century: America's Resurgence in the Asian Era*. New York: Crown.

Krugman, Paul R., and George N. Hatsopoulos (1987) "The Problem of U.S. Competitiveness in Manufacturing," in *New England Economic Review*, (January/February), pp. 18–29.

Kuznets, Simon (1966) *Modern Economic Growth: Rates, Structure and Spread*. New Haven: Yale University Press.

Lall, Sanjaya, ed. (1984) "Exports of Technology by Newly Industrializing Countries," *World Development* (May/June).

Lall, Sanjaya, and Paul P. Streeten (1977) *Foreign Investment, Transnationals and Developing Countries*. London: Macmillan.

Lam, Richard (1988) "Crisis: The Uncompetitive Society," in M. K. Starr, ed., *Global Competitiveness*. New York: W. W. Norton, pp. 1–42.

Lawrence, Robert A. (1984) *Can America Compete?* Washington, DC: Brookings.

Lima, J. L. Caldas (1984) *Monitoring of Technology Transfer Agreements by Regulatory Agencies: An Overview of Policies and Issues*. Geneva: UNCTAD.

Lodge, George C. (1984) *The American Disease*. New York: Alfred A. Knopf.

Lovell, Enid B. (1969) *Appraising Foreign Licensing Performance*. New York: National Industrial Conference Board.

Lucas, Barbara A., and Stephen Freedman, eds. (1983) *Technology Choice and Change in Developing Countries*. Dublin: Tycooly International Publishing Ltd.

McCulloch, Rachel (1988) "The Challenge of U.S. Leadership in High Technology Industries," *NBER Working Paper #2513*. New York: NBER.

McGarrah, Robert E. (1987) "The Decline of U.S. Manufacturing: Cases and Remedies," *Business Horizons*, 30 (November–December): 59–67.

Mansfield, E., and A. Ramo (1980) "Technology Transfer to Overseas Subsidiaries by U.S. Based Firms," *Quarterly Journal of Economics* 4 (December), 737–50.

Mansfield, Edwin (1988) "The Speed and Cost of Industrial Innovation in Japan and the United States: External vs. Internal Technology," in *Management Science*, 34, no. 10 (October), pp. 1157–68.

Milke, J., and P. Weston, (1988) *US International Competitiveness Evolution or Revolution*. New York, Praeger.

Mowery, David C., ed. (1988) *International Collaborative Ventures in U.S. Manufacturing*. Cambridge: Ballinger.

——— (1989) "Collaborative Ventures Between U.S. and Foreign Manufacturing Firms," *Research Policy*, 18, no. 1 (February), 19–32.

Muroyama, Janet, and H. Guyford Stever, eds. (1988) *Globalization of Technology: International Perspectives*, (Washington, DC: National Academy Press).

Mytelka, Lynn K. (1979) *Regional Development on A Global Economy: The Multinational Corporation, Technology and Andean Integration*. New Haven: Yale University Press.

National Science Foundation (1988) *International Science and Technology Data Update: 1988* (NSF 89-307), Washington, DC.

National Research Council (1984) *The Race For The New Frontier: International Competition in Advance Technology*. Washington, DC: NRC.

Nye, Joseph (1988) "Short-Term Folly Not Long-Term Decline," *New Perspectives Quarterly* (Summer), pp. 33–35.

——— (1990) *Bound to Lead*. New York: Basic Books.

OECD (1981) *North/South Technology Transfer: The Adjustments Ahead*. Paris: Organization for Economic Cooperation and Development.

Office of Technology Assessment (1984) *Technology Transfer to the Middle East*. Washington, DC: US Congress.

Ohmae, Kenichi (1989) "The Global Logic of Strategic Alliances," *Harvard Business Review* (March-April), pp. 143–54.

OTA (1984) *Technology Transfer to the Middle East*. Washington, DC: Office of Technology Assessment, U. S. Congress.

Parry, Thomas G. (1988) "The Muiltinational Enterprise and Restrictive Conditions in International Technology Transfer: Some New Australian Evidence," *Journal of Industrial Economics* 36 no. 3 (March), 359–65.

People's Republic of China (1984) "Guidelines for Science and Technology," *Beijing Review* (January 16).

Perlmutter, Howard V., and David Heenan (1986) "Cooperate to Compete Globally," *Harvard Business Review* (March/April), pp. 135–52.

Perlmutter, Howard and Tagi Sagafi-nejad (1981a) *International Technology Transfer: Guidelines, Codes and A Muffled Quadrilogue*. New York: Pergamon Press.

———— (1981b) *International Technology Transfer: Codes, Guidelines, and a Muffled Quadrilogue*. Book #1 in the Technology Transfer Trilogy. New York: Pergamon Press.

President's Commission on Industrial Competitiveness (The Young Commission) (1985) *Global Competition: The New Reality*. Washington, DC: GPO.

President's Report to the Congress (annual) *Science, Technology and American Diplomacy: Report submitted to the Congress by the President*. Washington, DC: U.S. Congress.

Quinn, James Brian (1970) "Scientific or Technical Strategy at the National and Major Firm Level," in UNESCO's *Role of Science and Technology in Economic Development*. Paris: UNESCO.

Rafii, Farshad (1984) "Joint Ventures and Technology Transfer: The Case of Iran" in Stobaugh and Wells, pp. 203–37.

Reich, Robert B. (1987) "The Rise of Techno-Nationalism," *The Atlantic Monthly* (May), pp. 63–69.

Reidenberg, Joel (1988) "Information Property: Some Intellectual Property Aspects of the Global Information Economy," *Information Age*, 10, no. 1 (January), 3–12.

Robinson, Richard D. (1984) *Internationalization of Business*. Hinsdale, IL: Dryden Press.

Robinson, Richard D., ed. (1987) *Foreign Capital and Technology in China*. New York: Praeger.

———— (1988) *The International Transfer of Technology: Theory, Issues, and Practice*. Cambridge: Ballinger.

———— (1989) *Cases on International Technology Transfer*. Gig Harbor, WA: Hamlin Publications.

Rodrigues, Carl A. (1985) "A Process for Innovators in Developing Countries to Implement New Technology," *Columbia Journal of World Business* (Fall), pp. 21–28.

Root, Franklin, and Farok Contractor (1981) "Negotiating Compensation in International Licensing Agreements: Actual Practice versus a Normative Model," *Sloan Management Review* (Winter), pp. 56–57.

Rosenberg, Nathan, and Claudio Frischtak, eds. (1985) *International Technology Transfer: Concepts, Measurement and Comparisons*. New York: Praeger.

Rostow, W. W. (1980) *Why the Poor Get Richer and the Rich Slow Down*. Austin: University of Texas Press.

Rugman, Alan M., ed. (1983) *Multinationals and Technology transfer: The Canadian Experience*. New York: Praeger.

Sagafi-nejad, Tagi (1979) *Developmental Impact of Technology Transfer: Theory, Determinants, and Verifications from Iran, 1954–1974*. Ph. D. dissertation, University of Pennsylvania.

Sagafi-nejad, Tagi (1984) "Egypt," in Lall, pp. 567–73.

Sagafi-nejad, Tagi (1987) "International Technology Transfer Agreements," mimeo.

Sagafi-nejad, Tagi (1989) "RBPs, TNCs, and LDCs: Major Issues Surrounding the Debate on Restrictive Business Practices of Transnational Corporations in Less Developed Countries," presented at the annual meeting of the Academy of International Business, Singapore (November).

Sagafi-nejad, Tagi, and Robert Belfield (1980) *Transnational Corporations, Technology Transfer and Development: A Bibliographic Sourcebook*. Book #3 in the Technology Transfer Trilogy. New York: Pergamon Press.

Sagafi-nejad, Tagi, and John Burbridge (1990) "Competitiveness and Global Manufacturing," in Daniel Orne et al., eds., *Global Manufacturing* (Greenwich, CT: JAI Press, forthcoming).

Sagafi-nejad, Tagi, and Stephen P. Magee (1982) "Foreign Direct Investment Theory and Capital Theory Considerations in the Pricing of Technology: An Empirical Investigation," mimeo.

Sagafi-nejad, Tagi, and Marc Rubin (1987) "Technology Transfer to the People's Republic of China: Congruence and Discontinuity in Perspectives," *Business Journal*, Special Issue #1, pp. 10–17.

Sagafi-nejad, Tagi, Richard W. Moxon, and Howard V. Perlmutter, eds. (1981) *Controlling International Technology Transfer: Issues, Perspectives and Implications*. Book #2 in the Technology Transfer Trilogy. New York: Pergamon Press.

Sarr, M. K., ed. (1988) *Global Competitiveness*. New York: W. W. Norton.

Schwartz, J. E., and T. J. Volgy (1985) "The Myth of America's Economic Decline," *The Harvard Business Review* (September–October), pp. 93–107.

Scott, Bruce R., and George C. Lodge (1985) *US Competitiveness in the World Economy*. Cambridge, MA: Harvard Business School Press.

Servan-Schreiber, J. J. (1968) *The American Challenge*. New York: Atheneum.

Simon, Dennis Fred (1983) "Technology for China: Too Much Too Fast? *Technology Review* (October), pp. 39–50.

Spencer, Daniel L. (1970) *Technology Gap in Perspective*. New York: Spartan Books.

Stipp, David (1990) "Soviets to Develop Technology in West in Joint Venture with Arthur D. Little," *Wall Street Journal* (February 22).

Stobaugh, Robert, and Louis T. Wells, Jr., eds. (1984) *Technology Crossing Borders*. Boston: Harvard Business School Press.

Succar, Patricia (1987) "International Technology Transfer: A Model of Endogenous Technological Assimilation," *Journal of Development Economics* (Netherlands) 26, no. 2 (August), 373–95.

Suttmeier, Richard (1980) *Science, Technology and China's Drive for Modernization*. Stanford: Hoover Institution Press.

Thunman, Carl G. (1988) *Technology Licensing to Distant Markets: Interaction between Swedish and Indian Firms*. Stockholm: Almqvist & Wiksell International.

UNCTAD (1981) *The Set of Multilaterally Agreed Equitable Principles and Rules for the Control of Restrictive Business Practices*. New York: United Nations.

—— (1983) "Monitoring of Technology Transfer Agreements by Regulatory Agencies: An Overview of Policies and Issues." Geneva: UNCTAD.

—— (1988a) "The Dimension, Direction and Nature of Technology Flows, Particularly to Developing Countries, in a Changing World Economy: Recent Trends in International Technology Flows and their Implications for Development," TD/B/C.6/145, 18 August. Geneva: UNCTAD 18 August.

—— (1988b) "Technology-Related Policies and Legislation in a Changing Economic and Technological Environment," TD/B/C.6/146. Geneva: United Nations Conference on Trade and Development.

UNCTC (1987) *Transnational Corporations and Technology Transfer: Effect and Policy Issues*. New York: United Nations Centre on Transnational Corporations.

—— (1988) *Transnational Corporations in World Development: Trends and Prospects* (New York: United Nations Centre on Transnational Corporations).

UNIDO (1981) *Case Studies in the Acquisition of Technology (I)*. New York: United Nations.

U.S. Congress (1980) "Study of U.S. Competitiveness," Washington, DC: Economic and Trade Policy Analysis Subcommittee of the Trade Policy Staff Committee, July, mimeo.

U.S. Department of Labor (1980) *Report of the President on U.S. Competitiveness*. Washington, DC: U.S. Department of Labor.

Vernon, Raymond (1966) "International Investment and International trade in the Product Cycle," *Quarterly Journal of Economics* (May), pp. 190–207.

—— (ed.) (1970) *The Technology Factor in International Trade*. New York: NBER and Columbia University Press.

—— (1971) *Sovereignty at Bay: The Multinational Spread of U.S. Enterprise*. New York: Basic Books.

—— (1977) *Storm Over the Multinationals*. Cambridge, MA: Harvard University Press.

—— (1986) "Can U.S. Manufacturing Come Back?" *Harvard Business Review* (July–August), pp. 98–106.

Vernon, Raymond, and Debora L. Spar (1988) *Beyond Globalism: Remaking American Foreign Economic Policy*. New York: Free Press.

Vickery, Graham (1986) "Technology Transfer Revisited: Recent Trends and Developments," *Prometheus*, 4, no. 1 (June), 25–49.

Wang, Jian-Ye, and Magnus Blomstrom (1989) "Foreign Investment and Technology transfer: A Simple Model," *NBER Working Papers*, 2958 (May).

Weiss, Stanley A. (1988) "Lessons From the Rise and Fall," *Harvard Business Review*, 66 (March–April), pp. 24–26.

Williams, B. R. (1972) *Science and Technology in Economic Growth*. New York: John Wiley & Sons.

Wolff, Arthur (1989) "Technology Transfer to the People's Republic of China," in *International Journal of Technology Management* 4 (4, 5): 449–76.

Wriston, Walter B. (1988/89) "Technology and Sovereignty," *Foreign Affairs* (Winter), pp. 63–75.

Yeh, Ryh-song, and Tagi Sagafi-nejad (1987) "Organizational Characteristics of

American and Japanese Firms in Taiwan," *Academy of Management Best Papers Proceedings 1987*, pp. 111–15.

Young, John A. (1988) "Technology and Competitiveness: A Key to the Economic Future of the United States," *Science* 241, no. 4863 (July 15), 313–16.

Zahlan, A. B., ed. (1978) *Technology Transfer and Change in the Arab World.* Oxford: Pergamon Press.

Contributors

JOSEPH BATTAT is an institutional development officer, Foreign Investment Advisory Service (FIAS), IFC/MIGA, the World Bank Group.

JACK N. BEHRMAN is Luther Hodges Distinguished Professor International Business Administration, University of North Carolina, Chapel Hill.

MICHAEL H. BERNHART is an associate professor of management, School of Business and Public Administration, University of Puget Sound, Tacoma, Washington.

FAROK J. CONTRACTOR is associate professor of international business, Rutgers University, New Brunswick, New Jersey.

WILLIAM A. FISCHER is professor of business administration, University of North Carolina, Chapel Hill.

ISAIAH A. LITVAK is professor of international business, faculty of administrative studies, York University, Toronto, Canada.

BRUCE MORGAN is president of Bruce Morgan Associates, a Washington, D.C.-based consulting firm with a long operating history in the Middle East.

RICHARD D. ROBINSON is George Frederick Jewett Distinguished Professor of Business, University of Puget Sound, Tacoma, Washington.

TAGI SAGAFI-NEJAD is professor of management, School of Business and Management, Loyola University, Chicago, Illinois.

DENNIS F. SIMON is associate professor of international technology, Fletcher School of Law and Diplomacy, Tufts University, Medford, Massachusetts.

D. ELEANOR WESTNEY is professor of business, M.I.T. School of Management, Cambridge, Massachusetts.

WILLIAM F. YAGER is assistant professor, School of Business Administration, Pacific Lutheran University, Tacoma, Washington.